CAMBRIDGE LIBRARY COLLECTION

Books of enduring scholarly value

Earth Sciences

In the nineteenth century, geology emerged as a distinct academic discipline. It pointed the way towards the theory of evolution, as scientists including Gideon Mantell, Adam Sedgwick, Charles Lyell and Roderick Murchison began to use the evidence of minerals, rock formations and fossils to demonstrate that the earth was older by millions of years than the conventional, Bible-based wisdom had supposed. They argued convincingly that the climate, flora and fauna of the distant past could be deduced from geological evidence. Volcanic activity, the formation of mountains, and the action of glaciers and rivers, tides and ocean currents also became better understood. This series includes landmark publications by pioneers of the modern earth sciences, who advanced the scientific understanding of our planet and the processes by which it is constantly re-shaped.

Life, Letters, and Works of Louis Agassiz

On the basis of extensive material in the form of letters, pamphlets and the recollections of friends and contemporaries, Jules Marcou (1824–1898) tells the story of the life and work of Louis Agassiz in this two-volume work of 1896. The Swiss-born palaeontologist, glaciologist and zoologist (1807–1873) is regarded as one of the founding fathers of the modern American scientific tradition. Marcou, a fellow countryman and collaborator of Agassiz, does not attempt to conceal his high regard for the subject of his biography but does have 'in view the truth'. In a chronological narrative, Volume 2 tells of Agassiz' professorship at Harvard and the founding in 1859 of the Museum of Contemporary Zoology, where he remained as director until his death. Although Darwin believed the Swiss scientist's theory on parallelisms provided evidence for evolution, Agassiz was no evolutionist but saw the plan of God everywhere in nature.

Cambridge University Press has long been a pioneer in the reissuing of out-of-print titles from its own backlist, producing digital reprints of books that are still sought after by scholars and students but could not be reprinted economically using traditional technology. The Cambridge Library Collection extends this activity to a wider range of books which are still of importance to researchers and professionals, either for the source material they contain, or as landmarks in the history of their academic discipline.

Drawing from the world-renowned collections in the Cambridge University Library, and guided by the advice of experts in each subject area, Cambridge University Press is using state-of-the-art scanning machines in its own Printing House to capture the content of each book selected for inclusion. The files are processed to give a consistently clear, crisp image, and the books finished to the high quality standard for which the Press is recognised around the world. The latest print-on-demand technology ensures that the books will remain available indefinitely, and that orders for single or multiple copies can quickly be supplied.

The Cambridge Library Collection will bring back to life books of enduring scholarly value (including out-of-copyright works originally issued by other publishers) across a wide range of disciplines in the humanities and social sciences and in science and technology.

Life, Letters, and Works of Louis Agassiz

VOLUME 2

JULES MARCOU

CAMBRIDGE UNIVERSITY PRESS

Cambridge, New York, Melbourne, Madrid, Cape Town, Singapore,
São Paolo, Delhi, Dubai, Tokyo, Mexico City

Published in the United States of America by Cambridge University Press, New York

www.cambridge.org
Information on this title: www.cambridge.org/9781108072618

© in this compilation Cambridge University Press 2011

This edition first published 1896
This digitally printed version 2011

ISBN 978-1-108-07261-8 Paperback

LIFE, LETTERS, AND WORKS
OF LOUIS AGASSIZ

AGASSIZ IN HIS LIBRARY
1861

LIFE, LETTERS, AND WORKS

OF

LOUIS AGASSIZ

BY

JULES MARCOU

WITH ILLUSTRATIONS

Vol. II

New York

MACMILLAN AND CO.

AND LONDON

1896

Norwood Press
J. S. Cushing & Co. — Berwick & Smith
Norwood Mass. U.S.A.

CONTENTS.

CHAPTER XIII.

1847 (*continued*)–1849.

CHAPTER XIV.

1849 (*continued*)–1852.

CHAPTER XV.

1852 (*continued*)–1855.

CHAPTER XIX.

1858-1864 (*continued*).

CHAPTER XX.

1865-1867.

CHAPTER XXI.

1868-1870.

CHAPTER XXII.

1871-1872.

CHAPTER XXIII.

1873.

CHAPTER XXIV.

APPENDICES.

ILLUSTRATIONS.

VOL. II.

CHAPTER XIII.

1847 (*continued*)-1849.

AGASSIZ was a constant visitor to the great markets
at Boston, purchasing all the fishes, crustacea, or wild
game he could find; and he was soon a great favourite
with all the market-men. What astonished them most
was his never satisfied desire for more specimens. He
would collect ten, twenty, and fifty specimens of the
same species of fishes, and turtles by the hundreds.

At each city where he stopped he never failed to
go to the market and carefully examine the stands of
fishermen and poulterers. At New York, with the
help of his cousin, M. Auguste Mayor, in a few days,

he filled a large barrel with specimens. Charleston's market, even more than the Boston and New York markets, contributed largely to his collections, and specimens accumulated rapidly. Barrel after barrel was filled to its utmost capacity; and Pourtalès and Girard had enough to do to keep pace with the professor's well-known propensity to get hold of every object of natural history with which he came in contact. The collections were at first placed in the upper story, or attic, of Tremont Temple, at Boston; but during the summer of 1847, they were divided into four parts: one of which was sent to Berlin; a second, to Neuchâtel; a third, to Paris; while the fourth was kept by Agassiz, and transferred to his house at East Boston, opposite Bird's Island, or placed in a wooden shed in the garden, where several tanks containing living animals constituted the aquatic laboratory.

At the request of the Faculty of the College of Physicians and Surgeons, of New York, Agassiz delivered in the hall of that institution, during the months of October and November, 1847, a series of twelve lectures, the full reports of which, as given in the columns of the "New York Tribune," widely attracted the attention of the American public. It was the first time that Agassiz's lectures were reported by stenographers, and printed in full, immediately after their delivery, and he was amused by hearing the newsboys in the streets of New York crying at the top of their voices, "Professor's Agassiz Lecture!" The demand for the papers containing these admirable discourses was so great, that the editor of the "Tribune" was

obliged to issue them in the form of a pamphlet, under
the title, "An Introduction to the Study of Natural
History.   Also, a Biographical Notice of the Author"
(New York, Dec. 10, 1847; Greeley & McElrath, Tribune Buildings).

His large and attentive audiences were so pleased
that the medical students and the New York doctors,
headed by the ornithologist, Dr. Trudeau, a special
friend of Agassiz, raised a subscription, which filled
a large box with silver dollars, and came in a body
to Dr. Trudeau's house, where Agassiz was a guest,
to present the offering to the professor, as a contribution, they said, toward the payment of the debts
contracted on account of his magnificent work on the
"Poissons fossiles."   Agassiz was much touched, and
tears sprang to his eyes, when, turning toward Dr.
Trudeau, who made the presentation, he thanked the
physicians, surgeons, and medical students of the great
city of New York for their very welcome and generous
gift.

At the end of one of these lectures Agassiz received
a visit, which was to him a complete and agreeable
surprise, from Jacques Burkhardt, of Neuchâtel, a
"salle d'armes" acquaintance as far back as his student
life at Munich.   After a rather uncertain life as an
artist in Rome, Burkhardt had returned to Neuchâtel just at the time that Agassiz was appointed professor; and as Agassiz was always in want of artists,
notwithstanding that he already had two — Dinkel and
Weber — in his service, he often employed Burkhardt
to draw fishes, and even took him, in 1842, to the

" Hôtel des Neuchâtelois," on the glacier of the Aar. Unsuccessful as a painter, notwithstanding the good teaching and advice of the two great Neuchâtel artists, — the brothers, Leopold and Aurel Robert, — Burkhardt, who was not an exact draughtsman, being skilful only in colouring the drawings after they had been made by others, led at Neuchâtel a precarious and unsettled life, and was unable to make both ends meet.

Disappointed in his endeavour to become an artist of repute, Burkhardt enlisted in a sort of half-military, half-colonial organization, created by the Belgian government as a means of establishing a colony in the district of St. Thomas, in Guatemala. When on the point of sailing, the Belgian government received a very strong protest from the president of the republic of Guatemala, against the sending of any military organization ; which he declared would not be accepted under any pretext. This put an end to the scheme, and left Belgium much embarrassed by the crowd of adventurers who had gathered ; and in order to dispose of them in the best manner possible, they were ordered to embark on a ship at Antwerp, with sealed orders, to be opened by the captain of the ship and the head of the expedition, when at a distance of several hundred miles from the Belgian port. The sealed orders directed the vessel to go to New York, and there to disembark the colonists, who were to be marched off to the Belgian consulate, and receive two months' pay, after which they were to be disbanded, and to go and do what they pleased.

Thus Burkhardt found himself in the streets of New York, ignorant of English, with only a small sum of money in his pocket, and without an acquaintance. He was, however, soon helped by a Neuchâtel merchant, established at New York, — M. Diacon, — and succeeded in making a humble living by drawing pictures on shades, and washing and mending old oil paintings.

Agassiz was in complete ignorance of what had become of Burkhardt since he left the glacier of the Aar and Neuchâtel in 1843. But Burkhardt, learning through the newspapers that Agassiz was delivering a course of lectures, gladly called on him and told his pathetic history, his attitude and appearance amply proving that life in the streets of New York under such conditions was a difficult one to endure. Agassiz, always open-handed and generous, received his old artist with great kindness, and offered him a home, on the single condition that he should draw his zoölogical specimens. The unfortunate artist was only too ready to accept any offer, or, more correctly, any arrangement, which promised a living; and with that lack of specific agreement which always characterized Agassiz's connection with his assistants, he resumed his position as draughtsman, and was brought by Agassiz to East Boston, on his return from New York, at the end of November, 1847. This hap-hazard association lasted until the death of Burkhardt, and is the only one, of all those formed in the same way, during the life of Agassiz, which remained undisturbed.

As soon as he had returned to Boston, Agassiz delivered another course of lectures before the Lowell Institute. His success increased with his facility in the use of English. He had entirely conquered American audiences, and his popularity grew at a pace which much astonished him. One day in January, 1848, he was approached by some of his friends,— among them most particularly Mr. John A. Lowell,— to know if a permanent professorship at Harvard College would be acceptable. His answer at first was a little hesitating; but the breaking out of the French Revolution in February, 1848, and the consequent great commotion all over the continent of Europe, including a revolution at Neuchâtel and rioting in Berlin, removed all his doubts; and he accepted the chair of zoölogy and geology, established specially for him by Mr. Abbott Lawrence, a Boston gentleman, who at this time founded the Lawrence Scientific School, in direct connection with Harvard University.

Never was a more happy appointment made at Harvard; it was a red letter day for the old university; for not only did Agassiz bring with him his unique reputation as a great naturalist, but his example of originating and urging forward new projects soon revolutionized the whole institution. Indeed, no one did so much to prepare for the new era of prosperity, and to increase the facilities of instruction, now so successfully organized and maintained under the presidency of Dr. Charles W. Eliot. But further, his children have since become the greatest patrons and benefactors of the university; for, taken altogether, they have already given not far from one million dollars,— the largest amount received from one family.

After his acceptance of the Harvard professorship, Agassiz, with Pourtalès, sailed for Charleston, South

Carolina, delivering there another course of lectures and continuing to collect specimens and make observations on the fauna.

The ease with which money could be made by public lectures rapidly turned the heads of Agassiz and all his household. His secretary Desor sent money to a German cousin, a gardener, asking him to come over at once, which he accordingly did. Then Desor arranged to have a regular emigration of assistants and attendants of all sorts from Neuchâtel to Cambridge, in order to make a permanent and large establishment. In this way an excellent lithographer, A. Sonrel, with the complete equipment of a designer and a printer, was secured. It was also decided to remove Agassiz's great library, and an order to pack up and to accompany it to America was sent to the librarian in charge, Henri Hüber; and finally two Swiss servants were also sent for.

During the absence of Agassiz in South Carolina, Desor, with the help of Dr. A. A. Gould, the learned conchologist of Boston, worked at a text-book of zoölogy, and as the book was to be printed in English, and as soon as possible, Desor lost no time in increasing his incomplete knowledge of English.

Before leaving for Charleston, Agassiz had rented a wooden house just built at Cambridge, the third house on the right side of Oxford Street, near the university; a much smaller house than the one at East Boston, and cheaper, costing only four hundred dollars a year; while the university had also secured for Agassiz's laboratory a small old bath-house close by Charles River, for the

storage of his collections and microscopical and anatomical studies. The removal was mainly effected during April by Pourtalès and Girard, who brought in a
dory all the specimens from the East Boston house,
and stored them partly in the cellar of Harvard Hall,
partly in the old bath-house near the Charles River,
and partly in the cellar of the Oxford Street house; and
when Agassiz returned from the South the 4th of April,
1848, he settled at once at Cambridge. Under some
pretext Desor had remained at the house at East Boston,
and it was even determined to keep that house a year
longer, in order to use it for the reception of all the
assistants and friends expected to arrive soon from
Switzerland, the Cambridge house being too small to
admit of such an increase of inhabitants. John A.
Lowell was frightened when Agassiz told him of what
he proposed to do, and it was with some difficulty that
he at last persuaded Agassiz to abandon the scheme,
as too expensive and entirely disproportionate to his
pecuniary resources.

    We have now arrived at the most critical moment in
the life of Agassiz. Different publications in French
and in German made by Desor and Karl Vogt are so
one-sided and ill-natured in their tone, that an exact
history of the whole affair is here an absolute necessity. The continual painful discussions, on scientific
and domestic subjects, between Agassiz and his secretary Desor increased to such an extent that Agassiz's
best friends on this side of the Atlantic, Messrs. Mayor
and Christinat, advised a separation.

Having just arrived at New York, at the beginning of May, 1848, I received, through A. Mayor,
an urgent invitation from Agassiz, to visit him and
spend a few weeks at his house in Cambridge, preparatory to an extensive tour in his company to
Lake Superior. I had been only a few days at Cambridge when I found myself involved in the turmoil
of the personal difficulties between Agassiz and his
secretary. The first who spoke to me of the matter
was Desor, who endeavoured to prejudice me against
Agassiz, and succeeded to a certain extent. However, from the start it did not please me that a man
who, ten years before, had come to Neuchâtel, unknown and without means, should speak so harshly
and so inconsiderately of the person who had received him more than kindly, had made him an inmate of his household, given him every opportunity
to rise in the world, and even taught him natural
history. Agassiz saw at once that I was influenced
to some extent by his secretary, and invited me to
a private talk. There he unburdened his heart,
sometimes sobbing and crying like a child. It was
extremely painful to him to be so ill-treated by one
who owed everything to him, although he was much
attached to Desor, whose qualities as a secretary and
assistant were highly praised and valued by him. At
the end of our long talk, Agassiz declared that he
would have no further connection of any sort with
Desor, and begged me, as the greatest favour he had
ever asked any one, to go to East Boston and tell Desor

that all relations between them were at an end, and
request him to vacate the house by the first of June,
when the lease would expire. I declined to give an
answer at once, saying that in twenty-four hours I
would make a reply.

My first impulse was to be out of the way, for I was
very much frightened by the responsibility and the awk-
ward position in which I was placed. Knowing no one
but the inmates of Agassiz's household, I asked advice
from every one of them. All denied Desor's accusations
as untrue, and all disapproved his conduct towards Agassiz. Christinat, who I knew represented the mother of
Agassiz, took me aside and insisted in the most positive
terms on their complete separation. He would not allow
compromise of any sort; and he insisted more especially
on the immovable determination of Agassiz's mother,
not to permit one of Agassiz's children to join him in
America, so long as Desor remained in the house.
This part of the information derived from Christinat
was decisive for me. However, I wanted Agassiz's
own words in regard to his children; for Desor had
repeatedly said in my presence that Agassiz did not
care anything about them. Agassiz was in much distress when he heard the accusation, and there was the
most pitiful scene imaginable. This idea of not seeing
his children around him again was so terrible that he
almost fainted away. Next day, after a sleepless night,
my decision was made; I had chosen to side with the
father, wife, and children against the adventurer, introduced in a fatal moment into the Agassiz household;
and I told Agassiz that I accepted his mission, however

disagreeable it might be, but that I wished to be accompanied by Frank de Pourtalès. The latter consented, and we went directly to the house at East Boston, where I delivered my message. Desor was at first stunned by it, but he soon recovered, and became insolent to such an extent that I withdrew, in company with Pourtalès, and we returned to Cambridge.

There Agassiz, moved to tears, took me by both hands and kissed me in the old Swiss fashion. He was full of thanks and compliments. He felt himself another man, because he had been relieved of a constant burden in his social and even mental life. For little by little Desor had taken such a hold on him, that he was not even free to express all his opinions and views on scientific subjects. In fact, he was controlled by Desor as by a manager, and not always considerately, being too often handled rather rudely. He had to provide all the money, and instead of being thanked for it, he was subjected to all sorts of moral tortures.

Matters took such a turn that friends interfered, and by common consent the whole difficulty was submitted to arbitrators. Agassiz chose John A. Lowell, Desor took Dr. D. Humphreys Storer, and the two elected as umpire Thomas B. Curtis, all of them among the first men in the city of Boston. After a thorough investigation the three arbitrators came unanimously to the opinion that Agassiz had been wronged by Desor, and consequently gave an award entirely in favour of Agassiz.[1]

So ended the scientific, social, and friendly relations between Agassiz and his German secretary, after a con-

[1] See the award, in "Trial of Desor *versus* Davis," pp. 53–56. Boston, Stacy & Richardson, printers, 11 Milk Street. 1852.

tinuance of about ten years, during the last three of which they were often turbulent and even violent.

Desor, after all his accusations against the man who had made him what he was, was bold enough to pretend that he had remained silent, and had only threatened to expose Agassiz; as if he had not attacked him in every way, both verbally and in print. (See "Synopsis des Echinides fossiles," par E. Desor, p. xx. Paris, 1858.)

The intervention of Professor Karl Vogt, an honest, but not always very exact and well-informed man, rather inclined by his eager disposition to see only one side of things, and to turn into ridicule every other view and opinion, has led me to give the real facts of the case, although I pass over many details. How Vogt in his biography of Desor ("Eduard Desor, Lebensbild eines Naturforschers," von Karl Vogt, in Deutsche Bücherei, in Zwanglosen Helten. IV. Serie, Heft 24. Breslau) could have declared that the award was in favour of Desor, it is difficult to understand, except on the supposition that he never saw the paper either in manuscript or printed, and was deceived by some one.

The scheme which he had prepared was an utter failure, happily for the natural history of America. He hoped to oblige Agassiz to leave Cambridge, and even the United States, when he meant to step in and take his place both officially and socially. Desor thought highly of himself, and over-shot his mark. He never was more than a second or third-rate naturalist, at the best, unable to go out of the beaten paths opened to him by Agassiz, Gressly, and Keller. He had no originality whatever, and seemed never to realize that, after all, he was only small change in comparison with the splendid medal "fleur de coin" of Louis Agassiz. His real value was quickly seen by Mr. John A. Lowell, who did not hesitate to uphold Agassiz, and never invited Desor to deliver a course of lectures at his institute, notwithstanding the pressure brought to bear on him by several friends of Desor, among them the celebrated Unitarian minister, Theodore Parker.

A few words more will dispose of Desor's doings in America. After receiving more than hospitality on board the *Bibb*, — for Lieutenant Charles H. Davis gave him pay as his secretary under the designation of master's mate, — Desor sued Commander Davis in December, 1851, before the Circuit Court of the United States for the district of Massachusetts, for breach of contract to write a memoir on the geological effects of the tidal currents of the ocean.[1] The jury gave a verdict for

[1] In a letter to Desor, dated February, 1849, written after the award between Desor and Agassiz, Davis said, "It appears from the award of the arbitrators, which, as bound to do both by honor and judgment, I fully accept — that in every point in dispute you have done injustice to M. Agassiz, and have misled those of your friends who were influenced by your representations"; and in another letter, dated March, 1849, he adds, "You speak of my condemning you unheard (it was M. Agassiz, if any one, whom I condemned unheard). . . . I have no wish to boast of any favors I may have conferred upon you. Nevertheless, I must say that the cordial and hospitable entertainment you received on board the *Bibb* last summer and autumn and my active but, as it appears, unavailing efforts to bring to a termination the unhealthy excitement under which your mind has labored towards M. Agassiz, may well relieve me from any painful sense of obligation to you." Finally, Admiral Davis wrote to Desor on the 5th of March, 1849: "No one can regret more sincerely than I do that a moral necessity, superior to all other considerations, has been created by yourself, which annuls these agreements" (*i.e.* the investigation of the subject of Ground Ice, the Natural Causes of Fogs on Shoals, and some other scientific topics, besides the Tidal Currents). — "Trial of Desor versus Davis," pp. 62-67, Boston, 1852.

the plaintiff, fixing the indemnity at a thousand dollars. The defendant moved for a new trial; and the case was definitely adjudged by his honour, Peleg Sprague, as follows: "The jury must have made a mistake. The verdict cannot stand in its present form. I shall give the plaintiff his election to remit five hundred dollars, or to take the opinion of another jury." The decision was accepted; ·and Lieutenant-commander Charles H. Davis was obliged to pay Desor five hundred dollars, besides his hospitality on board the United States steamer *Bibb*, his very generous treatment during Desor's stay, and his many acts of kindness during 1848.

After this performance, Desor, who had remained in America during 1849, 1850, and 1851, constantly causing as much annoyance as possible to his old benefactor and chief, had no alternative but to return to Europe, which he did in March, 1852, publishing, as a last Parthian arrow against Agassiz, his pamphlet, "Trial of the Action of Edward Desor, Plff., *versus* Charles H. Davis, Deft.," etc., Boston, 1852.

We have here a rare example of ingratitude in one who was elevated from nothing to a recognized place in the scientific world, and then turned against the hand which raised him from his obscurity and poverty.

Agassiz's first course of lectures at Harvard University was largely attended, not only by the regular students, but also by law students and several professors and instructors of the college and Scientific School. As soon as it was finished, Agassiz started to explore Lake Superior, accompanied by ten students, two gentlemen

from Boston, three doctors, and myself.   The rendez-
vous was a hotel at Albany, on the 15th of June, 1848.
On the same evening, Professor Agassiz began his daily
remarks on the region travelled over during the day,
giving a sort of itinerary lecture.   He had brought with
him a piece of black canvas and some chalk, and delivered a regular address, on rocks polished and scratched
by old glaciers and erratic pebbles and boulders, trans-
ported at a very remote epoch, and called attention to
the deposits of the red rocks of the Connecticut valley,
as well as to the vegetation of Massachusetts.

It was a very original and unique summer natural
history school; for Agassiz never repeated it, although
he said emphatically that he would do so every summer.
But circumstances were stronger than his desire; and
with the exception of a rather limited excursion to the
Adirondacks, the Lake Superior expedition remains his
only scientific exploration into the interior of North
America.   To be sure, he made several other explorations on the Atlantic coast, in Maine, the White Hills,
Massachusetts, South Carolina, and Florida, and created
a summer school at the island of Penekeese, as we shall
see; but he never made another scientific tour similar
to the Lake Superior excursion of 1848.

The tour extended through Niagara Falls, Lakes Erie
and Huron, with researches on the islands of Mackinaw
and St. Joseph, at the rapids of Sault Ste. Marie, and on
the whole northern shore of Lake Superior.   In birch-
bark canoes, containing three, four, or even five persons, besides three boatmen each, every feature of the
unsafe and sometimes dangerous shores was explored,

with halts at every interesting place, from Gros-Cap at
the outlet of Lake Superior to Michipicoten, Pic, and
Fort William factories of the Hudson Bay Company.
At that time the northern part of Lake Superior was a
perfect wilderness, all activity and marks of civilization
being confined to Point Keewenaw and its copper mines.
During the journey, which lasted from the 30th of June
until the 15th of August, only a few Indians, called
"gens du Lac" by the French Canadians, a branch
of the Ojibwa tribe, were met. The expedition when
at Fort William ascended the Kaministiquia River
as far as Kakabeka Falls, a distance of twenty-five
miles. There it separated; one canoe, the largest, containing five members of the expedition, myself among
them, left the main party, and started on the 25th of
July to make the round of the lake, returning to Sault
Ste. Marie by the south shore, in order to visit some
of the celebrated copper mines. Agassiz, with the rest
of the party, returned to the entrance of Lake Superior,
where he arrived the 15th of August, returning by the
same road.

The main results of the exploration were, first, an
extension of the glacial theory of Agassiz to include
all the shores of Lake Superior, almost an inland sea;
second, a thorough knowledge of the fishes of Lake
Superior and their comparison with those of the other
great Canadian lakes; third, a comparison of the vegetation of the northern shores of Lake Superior with
that of the Alps and the Jura Mountains; and fourth,
large collections of fishes, reptiles, birds, shells, and
insects, rocks, minerals, and fossils.

A remarkable volume entitled "Lake Superior: its Physical Character, Vegetation and Animals, compared with those of Other and Similar Regions," by Louis Agassiz, with a narrative of the tour by J. Elliot Cabot, and contributions by other scientific men, elegantly illustrated, appeared in due time, — March, 1850, — at Boston. A few words are necessary to call attention to the great value of the volume, which marked an epoch in natural history publications in America. Until then, all books containing plates of natural history objects, with a few exceptions, such as Isaac Lea's "Contributions to Geology," Conrad's "Fossil Shells of the Tertiary," "Natural History of New York," and Wilkes's "United States Exploring Expedition," had been executed in very poor style. Compare, for instance, the volumes of General J. C. Fremont, issued in 1845, with that of Captain H. Stansbury, issued in 1852, two years after the appearance of Agassiz's work on Lake Superior, both of which were published at the expense of the government. Stansbury's survey of the Great Salt Lake is in every respect a very creditable publication; while, on the contrary, Fremont's first and second expedition to the Rocky Mountains, Oregon, and Northern California is a disgrace as regards the execution of landscapes and natural history illustrations. Agassiz, who had succeeded in bringing the lithographic establishment of M. A. Sonrel from Neuchâtel to Cambridge, put into the hands of that excellent French artist all the illustrations and drawings of the landscapes and specimens. Everything was done in fine style; and the volume, when published, attracted attention, and even admiration; the best proof of its great value being the fact that now it commands more than double its original price.

Sad, but not unexpected, news awaited Agassiz on his arrival at Cambridge, on the 28th of August. The death of his wife, at Freiburg-im-Breisgau, Grand Duchy of Baden, had occurred on the 27th of July, 1848, two days before her thirty-ninth birthday. She had learned, with great relief of mind, the separation of Agassiz from his secretary; but she was too ill to hope for recovery, for consumption had set in. After a long struggle she died, surrounded by her three children, in the house of her beloved brother Alexander Braun, professor of botany at the University of Freiburg,[1] and was buried at the old cemetery of that town, closed since 1867, and no longer used for burials. A simple granite gravestone, with only her initial letters, C. A. (Cecile Agassiz), engraved on it, marks her resting-place; no date, no sign, indicates that here reposes

---

[1] The accomplished daughter of Alexander Braun, Mrs. Cecile Mettenius, says in the life of her father : " After many days of suffering, during which hopes of recovery mingled with presentiments of the approach of death, Mrs. Agassiz died quietly, the 27th of July, 1848. A few days before her death, she received with joy a letter from her husband, to whom she had sent the portraits of her children, drawn during her sickness." It was a last tribute of love. Mrs. Mettenius adds : "She had had a life full of struggles and of sorrows." Braun wrote to his brother (Max Braun), announcing the sad news: "Our sister, who has had so many afflictions, has found to-day her rest after her stormy life. She has suffered much. God will give her in the other life what will change to joy all the suffering of the life on earth; there she will understand the divine Providence which is full of charity, but whose ways are so obscure to us." ("Alexander Braun's Leben," p. 405; Berlin, 1882.)

the first wife of the great naturalist, Louis Agassiz, the mother of his children.

Alexander Braun had removed his family from Carls- ruhe to Freiburg, after his appointment as professor of botany at the university; and almost as soon as he was settled there, his sister, Mrs. Agassiz, became so ill, that, after the month of December, 1847, she was unable to leave her bed, except for a few hours each day. Phthisis made rapid progress and became incurable.

Professor Braun kept with him the children of his sister until he heard from Agassiz, who asked him to conduct his two daughters to Switzerland, to their grandmother, where they were to remain until he could himself go for them, or arrange for their joining him in his lately adopted country. The son Alexander, then twelve years old, stayed a year longer in the Braun family. Before separating, the children received the first visit from an American family, Mrs. Bruen and her two daughters, the oldest since so well known as the wife of the celebrated author of "The Italian Sculptors," Charles C. Perkins, of Boston, friends of their father, who came specially to Freiburg, in August, 1848, in order to tell them, *viva voce*, how kindly their father had been received, and how highly he was esteemed in the New World. A few weeks after this visit, Braun took his two nieces, Ida and Pauline, and placed them under the guardianship of their grandmother, Mrs. Agassiz, at Cudrefin, on the lake of Neuchâtel.

The year 1848 was most eventful in the life of Agas- siz: first, in his appointment as professor at Cambridge; second, in the dismissal of his secretary, Desor; third, in the death of his first wife. The first two events proved most beneficial to his future life.

Agassiz found at his house M. Sonrel and his wife, awaiting his return from Lake Superior. A house was immediately leased in the vicinity, and Sonrel began in earnest the establishment of his lithography. During September his two workmen arrived, a draughtsman and a pressman. Then came, in succession, Professor Arnold Guyot, with a nephew and a cousin; Charles Girard's brother and sister; Hüber, with the library and some specimens of rocks and fossils; and M. Leo Lesquereux, with his wife and five children. Every one was lodged, at least, for several days, in Agassiz's Oxford Street house. Mattresses were laid on the floors of different rooms, even in the parlour; the only unoccupied room being the dining-room, where the table was always abundantly furnished. It was a second "Hôtel des Neuchâtelois," transferred from the glaciers of the Aar to Cambridge. In all, there were twenty-three persons, twenty-two of whom came from Neuchâtel, town or canton. Pourtalès and Marcou soon left, but were replaced at the dining-table by the two German assistant chemists of Professor Horsford. In fact, it was "la Maison du Bon Dieu," as the French call it, every one entering any room, and Agassiz receiving with a smile every new arrival.

Some details may be interesting, for it is not likely that such a naturalist household will ever be seen again. "Papa" Christinat, as he was called, was the general manager. More than sixty years of age, he was still extremely active, possessed excellent health, and was most devoted to the welfare and interest of Agassiz. Although not speaking or understanding more than half a dozen English words, he had the skill to make himself understood, by gestures and pantomime, which he executed with great readiness, and generally with good results. He succeeded in teaching the cook, an illiterate Irish girl, a sufficient number of French words to make her understand his directions in regard to some French dish to his taste. His most remarkable performance was the buying of provisions. Not satisfied with the butcher who brought meat and vegetables to the door every day, he concluded to go himself to the great markets in Boston. Every day, he walked into Boston, starting at daybreak; and, to economize, he carried a great basket on his arm, and so calculated his time, as to arrive at the Cambridge bridge before the toll office on the Cambridge side was open. Then he would stand on the bridge, between the two toll houses, for five, ten, or even twenty minutes, waiting patiently until the toll keeper on the Cambridge side opened his small cabin and office, — an indication that the toll had been paid; and then Christinat would pass in front of the toll house on the Boston side, thus avoiding the toll of one cent!

At the Faneuil Hall and Quincy markets, he always made excellent choice of meat, fowls, fishes, lobsters, and vegetables, for he was truly "a connoisseur"; but he spoke French to all the market men, his only English being *how much?* and he never understood the answer; but paid what he thought was proper. The men almost always remonstrated on the ground that it was not the

price asked for. "C'est assez!" was the invariable
answer of M. Christinat, who then triumphantly put the
purchase into his basket, marching off as if he were perfectly satisfied. As he was known all over the market
as the friend and housekeeper of Professor Agassiz, the
traders let him do what he pleased, but marked the difference of price in their books. At dinner, Papa Christinat
would say : " Comment trouvez-vous le gigot de mouton ?
ou le poisson ? " " Excellent ! " was the ready answer of
all the guests ; " Eh bien ! je ne l'ai payé qu'un dollar et
quart," when the piece, according to the market value,
may have been two dollars or two dollars and a half.
Agassiz used to smile quietly and compliment Christinat on his fine bargain. But, alas ! at the end of the
month came long bills from the fisherman, the butcher,
and provision dealer, which put an end to Christinat's
method of purchasing at markets at low prices.

All sorts of specimens in alcohol, and even alive, were
constantly coming by express, sent from all parts of
New England and even from more distant parts of the
country. One day, to the great amazement and amuse-
ment of all the inmates of Agassiz's American " Hôtel
des Neuchâtelois," the express brought a live big black
bear sent from the forests of Maine by an admirer of
Agassiz's lectures at the Lowell Institute. There was
no other indication on the label attached to the neck of
the animal. This time the professor had " caught a
Tartar " ; he had no place to keep it, and he begged an
expressman to keep it in his stable. There, all the
horses were made uneasy by such a neighbour, and
after a few days the animal was disposed of by poison-

ing, and dissected, his bones being preserved for the collection.

Another day there arrived, on a wagon, a single specimen, so large and weighty, that it was as much as the horse could do to drag the heavy load. It was an enormous sea-turtle, the gigantic leather-back (*Sphargis corriacea*), found cast up on the coast of Cape Cod, after a violent storm. The big animal, an inhabitant of the Caribbean Sea, had lost one of its largest paddles, near his head, cut off close to the carapace, very likely by a shark; and although the wound was well healed, the loss of the paddle incapacitated him for facing a Gulf Stream storm, which carried him far north, until beached on the sand of Cape Cod. He was carried to the old bath-house, where dissection was begun. Christinat, with an eye to economy, noting the beautiful veal-like meat, took some home, and at dinner it was so praised and relished, that for a week the numerous guests of the " Hôtel des Neuchâtelois " lived on turtle soup, turtle steak, turtle pie, and turtle roast.

It was really beautiful to see Agassiz struggle to provide for all the expenses of his household. In the three months from September to December, 1848, he had spent three thousand dollars, while his salary at the Lawrence Scientific School was only sixteen hundred dollars a year. John A. Lowell came forward at once, paying him twelve hundred dollars in advance for another course of lectures. But even this was not enough, and Agassiz had to accept every offer made to him for lectures in the towns and villages around Boston. Besides his two regular lectures every week at Harvard University, and his two Lowell lectures, Agassiz went every afternoon, when not engaged to the Lowell Institute, to some suburban town, like Salem, Framingham, Worcester, etc., and delivered an evening lecture, accepting any offer, however small, which was offered to him. Only his strong constitution enabled him to stand such fatigue; for, as too often happened with him, he was burning the candle at both ends. However, he was happy, for his success as a lecturer was beyond his most sanguine expectation; his audiences being so large that he was obliged to repeat next day each of his lectures at the Lowell Institute. "The Evening Traveller," a Boston newspaper, similar to "The Tribune" of New York, had each lecture stenographically reported, and published it with woodcuts the day after its delivery, and the sale was so great that the newspaper was obliged to reprint each number containing these lectures, and finally to issue them in the form of a pamphlet, under the title, "Twelve Lectures on Comparative Embryology, delivered before the Lowell Institute, December and January, 1848-49," Boston, 8vo. The newsboys in the streets of Boston and Cambridge used to cry Professor Agassiz's lectures at the same time that they announced a revolution in Europe, or a shipwreck of a great transatlantic steamer, or the election of General Taylor as President of the Republic. No one was more popular in New England than Agassiz; he even rivalled the great statesman, Daniel Webster.

During 1848 Agassiz's prominent part on two public scientific occasions showed what a high place he already held in America as a scientific leader. On May 17 the Boston Society of Natural History dedicated a new building in Phillips Place, at which time the Annual Address was delivered by Dr. D. H. Storer. The new building was very well suited in every respect to the wants of the society, and was filled to its utmost capacity, in expectation of hearing a speech by Professor Agassiz. At the conclusion of the vice-president's address, Agassiz made remarks on the future progress of the natural sciences in America, insisting on the excellent opportunity offered by the political disturbances in Europe, where progress was forcibly suspended for some time to come. If strenuous and sacrificing efforts are made now, he said, they will soon bring results which will place America in the position hitherto occupied by Europe. Tremendous applause from the great audience proved that the American scientific public highly appreciated Agassiz's endeavours to promote the cause of American science.

The second occasion was the ninth annual session of the Association of American Geologists, held the 20th of September at Philadelphia. An organic change in the name and purpose of the society had been proposed at its last meeting at Boston, the year previous, and an enlarged constitution embracing all sciences, somewhat on the same plan as that of the British Scientific Association for the Advancement of Sciences, had been framed by a committee of three, Agassiz being one, and we may say the leading member; for he had assisted at several meetings of the great British association in England, and had even presided over the Société Helvétique des

Sciences Naturelles, a precursor, and the first great European association to promote progress in natural and physical sciences, outside of scientific organizations existing permanently in large cities. Full advantage was taken of his experience in the new organization, and the Association of American Geologists and Naturalists became the American Association for the Advancement of Sciences.

CHAPTER XIV.

1849 (*continued*)–1852.

At the beginning of February, 1849, Agassiz left Cambridge for a prolonged visit to Philadelphia, and having agreed to deliver a course of lectures, he remained there until the middle of April. His success was so great that Philadelphia savants and society leaders approached him with a view to an appointment at the Pennsylvania University, but, as he said, it was too late; his decision was made to remain in Cambridge. During the intervals of his lectures Agassiz was very busy at the Academy of Natural Sciences. The Academy was then the best equipped institution in America. Its museum contained a large number of typical specimens, fossil and living, described by Morton, Conrad, Say, Nuttall, Audubon, Lucien Charles Bonaparte, Harlan, Rafinesque, and others, besides a great quantity of inedited species, and its library was very rich and kept constantly *au courant*

27

through the liberality of one of the members of the Academy, Dr. Thomas B. Wilson. Agassiz had brought with him his artist Burkhardt and his librarian Hüber, and both had their hands more than full during their stay, for Agassiz's activity knew no bounds. There were at that time in Philadelphia a number of naturalists, the most prominent being Morton, Conrad, Lea, and young Leidy.

Dr. Samuel George Morton was the most remarkable American naturalist then living; unhappily, he was an invalid, suffering from a very serious affection of the heart from which he died two years later. Morton was the real founder of invertebrate palæontology in America. His volume entitled "Synopsis of the Organic Remains of the Cretaceous Group of the United States," 1834, is the starting-point of all palæontological and systematic work on American fossils. His "Crania Americana," 1839, and "Crania Ægyptiaca," 1844, placed him at the head of ethnologists in the Old and New Worlds. It was natural that a friendly understanding should promptly arise between Agassiz and himself, notwithstanding the sad condition of Morton's health. Agassiz became a strong advocate of the doctrine of different species of man; the word "race" being reserved, and applied only to varieties in each species; and, as a contribution to the ethnological researches of Morton, Agassiz wrote, after Morton's death, his celebrated "Sketch of the Natural Provinces of the Animal World and their Relation to the Different Types of Man," which began the first volume of "Types of Mankind" (1854) dedicated "to the memory of Morton."

Agassiz saw a great deal of Morton during his two months' stay at Philadelphia, frequently visiting him, for Morton was already confined almost constantly to his library. I may say that after George Cuvier, Morton was the only zoölogist who had any influence on Agassiz's mind and scientific opinions. Of course, I do not refer to the glacial question, which belongs to another order of studies entirely outside of pure zoölogy. I several times enjoyed the privilege of accompanying Agassiz on these visits, and was much impressed by his enthusiasm. He had, at last, found a naturalist to his liking, without any reserve.

Timothy A. Conrad, then curator of the museum of the Academy of Natural Sciences, was a palæontologist of the first order; but ill-health prevented him all his life from doing himself justice. Although his studies were limited to other departments of palæontology, Conrad was much attracted by the varied and profound knowledge of Agassiz. He delighted in showing him all the rare specimens contained in the museum, and finally he succeeded in organizing an excursion into New Jersey to Timber Creek and vicinity, in order to show the typical upper cretaceous of the Atlantic States. Agassiz was suffering at that time from overwork and anxieties of all kinds, and it was with difficulty that he was persuaded to be one of the party. On a beautiful early spring day in March, after awaiting at Camden the arrival of Agassiz, who was never an early riser, the party, composed of Conrad, Agassiz, Leidy, Dr. Hallowell, myself, and two of Agassiz's Cambridge students, started for Timber Creek, under the guidance of Dr. Harris, a

resident of that part of the country. We saw the celebrated New Jersey greensand, and collected in it a
quantity of sauroid teeth, fragments of vertebræ of a
crocodilian, *Ostrea vesicularis, Terebratula Sayii, Pecten,
Arca, Mya,* etc. Above the greensand and green marls
there are yellow sands and yellow calcareous sandstone
containing many echinoderms, such as *Holaster, Nucle-
olites, Diadema, Hemiaster, Ceratomus;* and many *Teredo,
Scalaria,* and corals; the whole formation belonging
to the lower chalk and upper greensand of England.
The excursion was very successful, and Agassiz returned
to Philadelphia in better health and spirits.

Dr. Joseph Leidy, so justly celebrated since for his
great works on the comparative anatomy of American
fossil vertebrates, was then a young student just returned
from Paris, where he had followed the lectures and also
the private instructions of de Blainville, the successor
of Cuvier in the chair of comparative anatomy. Leidy
was the naturalist who followed, with the greatest attention and best results, the lectures of Professor Agassiz,
and received the most benefit from his presence in
Philadelphia. Agassiz was much impressed by his
modest manner, and predicted a great future for him if
he would devote his life to comparative anatomy. Many
fêtes were given by Philadelphia society in honour of
Agassiz, and it was not without regret that on April 11
he left the city of Penn, returning directly to Cambridge.

Resuming his duties at Harvard University, he devoted himself, during the year, to microscopical studies,
more especially of the *Acalephæ* or *Medusæ* of the shores
of Massachusetts, which resulted in the publication of

a monograph in two parts, in the "Memoirs of the American Academy of Sciences," of which the sixteen quarto plates, drawn on stone from nature by Sonrel, are simply superb. This first contribution of Agassiz to the natural history of the United States marks a new departure in America, in that it shows the structure, the generic peculiarities, and all the characters of the muscular system in very low animals.

Agassiz gave a great part of his time also to directing the publication of his volume on Lake Superior. Curiously enough, he wholly abandoned his "Principles of Zoölogy." His collaborator, Dr. A. A. Gould, wrote, during 1849 and 1850, the second part, as he had already written with success the first part; but from year to year Agassiz delayed reading and correcting the manuscript, until it was too late to publish it. It was one of the foibles of the great naturalist not to finish promptly the work already begun, but to let it drop in order to undertake other work. However, the success of the first part on "Comparative Physiology," which was issued in 1848, was great, and several editions were printed. The work was quickly pirated in England by unscrupulous editors. A German translation by Professor Bronn, the celebrated palæontologist of the University of Heidelberg, was published at Stuttgart in 1851; and a French translation, by the great geographer Élisée Réclus, was printed in Paris, in the "Magasin d'Éducation et de Récréation," as late as 1891, under the title "Principes de Zoologie." The work had great influence in America, the three editions of 1848, 1851, and 1861 being extensively used by professors and students, and it is to be regretted that it was never completed.

A great pleasure was in reserve for Agassiz; namely, the arrival of his son Alexander in the middle of June, 1849, brought from Neuchâtel by a cousin, Dr. Mayor, and the evangelist, Marc Fivaz, of Newark Valley, in the state of New York, the first naturalist companion of Louis Agassiz at Orbe. Born at Neuchâtel the 1st of December, 1835, Alexander, as he was christened, in honour of his uncle, Alexander Braun, was a lad of thirteen years, well developed and fine looking, but more serious and inclined to solitude than boys of his age generally are. Agassiz was delighted and grateful for all the marks of interest and kindness shown his son by every one in Cambridge and Boston. In a letter to me, dated Cambridge, June 20, 1849, Agassiz says: " Je reviens de New York avec mon fils, c'est vous dire que je suis bien heureux maintenant. C'est dans toute la vérité de l'expression et à part de la partialité paternelle un charmant garçon."

From the first day of his settlement at Cambridge, Agassiz was befriended by both Professors C. C. Felton and Benjamin Pierce. Every day they called at his house, and generally more than once, helping and cheering him by true friendship. Professor Felton, whose second wife was the granddaughter of Colonel Perkins, — a merchant-prince of Boston of the beginning of this century, — had a very attractive home, in which Agassiz was always welcome, and even indulged to his heart's content. Felton was an extremely amiable man, and a scholar of repute ; and mutual attraction soon brought him and Agassiz together in an intimacy which lasted until the death of Felton in 1862.

Mrs. Felton saw quickly the influence she exercised over Agassiz, and used it for his advantage. Agassiz was always in need of good advice; for in almost all the walks of life outside of science he was like a child, completely lacking in knowledge of men and good judgment in matters of domestic life. He formed the habit of asking the advice of Professor and Mrs. Felton, and what is better, he followed it as far as it was possible with his enthusiastic nature.

When a youth in Germany, Switzerland, and Paris, Agassiz wore a heavy moustache, which he was obliged to cut rather reluctantly when appointed professor at Neuchâtel, where the society was then formal and conventional. During his exploration on Lake Superior he had let his moustaches grow, and came back to Cambridge with a formidable pair. Then moustaches were absolutely abhorred in America, so much so that I was insulted repeatedly in the streets of Buffalo, Cleveland, and Detroit, because I wore a light moustache and a light beard. As soon as Mrs. Felton saw Agassiz on his return, she had no difficulty in persuading him that a moustache was not becoming to him, which was true enough; and the next day Agassiz appeared completely shaved, with the exception of slight whiskers, which he wore until the end of his life.[1]

---

[1] It may seem too unimportant and even trivial to refer to such a small matter, but it shows the strong prejudice then existing in America against moustaches, a prejudice which was soon to disappear with the great Civil War.

Mrs. Felton's second sister, Miss Elizabeth C. Cary, had always felt a great admiration for the rare gifts of Agassiz as a public lecturer since his first course of lectures at the Lowell Institute, and as she was a constant visitor at her sister's house, a friendship soon sprang up between Agassiz and her which in due time on his part changed into courtship.

After the return of Agassiz from Philadelphia, the engagement was announced, the marriage taking place at King's Chapel early in the spring of 1850.

It was certainly the best thing Agassiz ever did in the course of his whole life.

However, at first the news of the engagement did not please Agassiz's old friend, M. Christinat, and his Swiss family. With Agassiz's habit of spending money lavishly in every direction, Christinat and Mrs. Agassiz, the mother, thought that if he married again he ought to marry a very rich wife, able to support a great establishment, as his salary would never be sufficient to carry out one-tenth part of the schemes of which his head was always full. Miss Cary had no fortune of her own, and Christinat, although very favourably impressed by the young lady, opposed the marriage, because he thought that instead of helping Agassiz, it would add a new burden and be the occasion of additional expense. As soon as he saw that the engagement was concluded, Christinat resolved to leave Cambridge, being unwilling to witness the marriage; and, therefore, in November, 1849, secretly, without a word to Agassiz, who was absent on a lecturing tour in Massachusetts, he left

Cambridge, embarking for New Orleans. He kept his own secret, and Agassiz did not know of his whereabouts until more than a˙ year later, when he learned that he had passed the year at New Orleans, as pastor of a Swiss church there. Agassiz, who knew the difficulty of his family's position, urged him to finish his life with his " viel ami," and when his fiancée was made acquainted with the arrangement, she cordially acquiesced; but it was all in vain. Christinat, to my great surprise, recognized me one day in a church in Paris early in January, 1851; he was returning to Switzerland, after a rather trying experience in Louisiana, where there had been a severe epidemic of yellow fever, from which he escaped almost miraculously. He was reinstalled as pastor over his old parish at Montpreveyres, Canton de Vaud, in April, 1852, and died there the 20th of February, 1855.

Instead of being a burden in her new home, Mrs. Agassiz was an invaluable addition, and a comfort to all. An excellent manager of her household, she at the same time undertook to act as secretary to her husband, always ready for any new task. It was, indeed, remarkable to see a young lady, brought up in great comfort and leisure, enter a life full of responsibilities of all kinds, even becoming an authoress in order to help and please her husband. She was and is still the guardian angel of Louis Agassiz and his whole family of children and grandchildren; and blessed was the day for Agassiz when she stepped into his house in Oxford Street.

Christinat was not the only inmate of the second
" Hôtel des Neuchâtelois " who left it during 1849 and
1850. First, Pourtalès received an appointment in the
United States Coast Survey, and became a resident
of Washington. Guyot brought over his rather large
family, and settled for a time in Cambridge. Lesquereux went West, and made his home at Columbus,
Ohio. Charles Girard went to Washington, as an assistant of Professor Spencer Baird, at the Smithsonian Institution; and, finally, the librarian, Hüber, returned to
Switzerland, at the end of 1849. To him is due, in a
great measure, the executive part of the " Bibliographia
Zoologiæ et Geologiæ," and to him and Girard together,
the rather difficult work of the " Nomenclator "; and
because of his scholarship and linguistic ability, he was
sent by Agassiz to Soleure, in 1846, to arrange for the
printing of the " Nomenclator."

At the end of August, 1850, Agassiz's two daughters
arrived, in the care of the good cousin, M. Auguste
Mayor, who was always ready to help his friends and
relatives, and all Louis Agassiz's family was at last
gathered under his roof. It was a great achievement,
and an immense relief to his mind, after many years
of anxiety and suffering, and a new life was now in
store for him.

Once surrounded by his children, Agassiz recollected
all his younger days and life in Switzerland; and Mrs.
Agassiz had the happy thought of putting on paper all
she heard, in the form of a journal, which she has since
used to much advantage in the first half of the first vol-

ume of Agassiz's Life. Agassiz corrected the manuscript; and it was occasionally read aloud to his children, as an exercise in their English studies.

It was no small task for Mrs. Agassiz to manage a family which, until then, had never used the English language, and whose manners and thoughts were those of the French Swiss, with many German elements. She applied herself with rare perseverance, much gentleness, constant watching, to Americanize the whole family. As a rule, she never spoke French, although understanding it perfectly. Her uniformly calm manner and temperament helped her immensely; she took everything quietly, never losing her temper, always serene, and determined to attach Agassiz and his children to America, at whatever cost.[1] She was admirably seconded by her sister, Mrs. Felton; and it is not too much to say that to these two ladies is due Agassiz's remaining in the New World and the Americanization of his children and grandchildren. At first Agassiz was easily won; he even affected, during the first twelve years of his union with his second wife, to see and hear nothing but what was American, severing almost every tie with Europe. If it had not been for the prolonged life of his mother, who died only seven years before him, he might have been considered a devoted naturalized American citizen. But a gradual change came

[1] There is a sentence taken from a French author, very applicable to the second marriage of Agassiz: " Il n'y a d'amitié durable et féconde qu'entre gens qui ne se ressemblent pas." How true of Agassiz and his second wife. They differed in every way: education, character, disposition, and ideas.

over him after his journey to Brazil; more and more
frequently he remembered his Swiss origin, and became
more and more attached to the French civilization.

Early in January, 1851, Agassiz started for an exploration of the Florida coral reef. His friend, Professor
Bache, the Superintendent of the Coast Survey, secured
his valuable services and put at his disposal and under
his orders the schooner *W. A. Graham,* of the Coast
Survey. During his ten weeks' survey he studied the
growth of coral reefs, their mode of living, the different forms and associations of animals in the reef and
around it; and the recent and present formation of
shell limestone and oölitic limestone; but he found no
traces of upheaval or subsidence, even at the Tortugas.

In August, 1851, Agassiz addressed his report to
Professor Bache, who published extracts from it in his
Annual Report, Washington, 1852, on pp. 145–160.
Bache was so impressed by the Report, that he asked
Congress for an appropriation to give the entire results
in detail, with drawings of all the species of corals; and
the specimens were put directly into Sonrel's hands, to
be drawn on stone from nature. The plates were all
struck in beautiful style and are works of art by themselves; they were even paid for by the United States
government; but this was all, for it was impossible to
get the text describing the species. I saw Agassiz,
just after his return from Florida, full of his subject,
enthusiastic, as he always was, urging M. Sonrel to
finish the plates as quickly as possible, and I certainly
thought that his full observations would be before the

scientific world in a year or two at most, and would give
new views on coral reefs, barrier reefs, and atolls,
entirely different from those presented by Darwin,
Couthouy, and Dana in describing those of the Pacific
Ocean.

Agassiz always had more irons in the fire than he
could manage, and instead of diminishing their number,
he was constantly increasing them. Happily, his son,
long after his death, took hold of the plates, reprinted
the whole of the Report sent to Professor Bache in
1851, and with the help of Pourtalès, who named all the
figures, issued in 1882, in Vol. VII. of the " Memoirs
of the Museum of Comparative Zoölogy," the final
" Report on the Florida Reefs," by Louis Agassiz, 4to,
Cambridge. The editor has added at the end of the
paper a chapter "entitled, Sketch of the Florida Reefs
and Keys," extracted from a small volume of Louis
Agassiz, entitled " Methods of Study in Natural History," Boston, 1863, 12mo.

The great number of specimens collected in Florida,
added to the already important collections gathered at his
house and at the old bath-house by the Charles River,
made it an absolute necessity to build a sort of laboratory
with a lecture-room, and a quantity of drawers for the
display of the specimens. A wooden structure, for the
storage of the specimens preserved in alcohol, was
therefore erected on the college grounds, to the left of
the chemical laboratory and engineer's room of the
Lawrence Scientific School, close by the present Hem-
enway Gymnasium, and was ready for occupancy in

the spring of 1852. It was only a temporary museum
— a rather dangerous wooden structure for the storage
of specimens kept in alcohol. The University voted a
sum of four hundred dollars annually for the preserva-
tion of the collections; and some friends started a subscription of twelve thousand dollars to purchase them
for the Lawrence Scientific School, as the expense
incurred by Agassiz was too great to be borne by a
man without a private fortune.

On a beautiful September afternoon of 1851, when
crossing that college yard now occupied by Memorial
Hall, I was stopped by a gentleman in full dress, —
frock coat and black pantaloons of an old pattern, and
too short by six inches, showing the upper part of his
boots, — who asked me, in broken English, to direct him
to M. Agassiz's house. "I am going there myself," I
answered in French. "Oh! vous êtes français; je l'ai
pensé en voyant votre barbiche et moustache, car depuis
mon arrivée à Boston, je n'ai vu personne que vous et
moi avec des moustaches." We walked along leisurely,
and I thought from his conversation that he was a
countryman from the Canton de Vaud — perhaps a rich
Swiss farmer. Agassiz was at home, and, on seeing my
companion, exclaimed: "Oh! M. Ampère!" In fact, it
was no less a person than the son of the great electri-
cian and physicist, André M. Ampère, himself a very
remarkable scholar and savant in his way, and distinguished in literature, and the last admirer and lover
of the beautiful Madame Recamier. His visit pleased
Agassiz much, showing as it did that he was not for-

gotten by his numerous Paris friends.   Both were great
and most agreeable talkers; and reminiscences of all
kinds so rapidly succeeded each other, and time passed
so quickly, that M. Ampère was persuaded to pass the
night at Agassiz's house.

It may be said that Ampère[1] was engaged, when a
young man, to marry the daughter and only child of
George Cuvier, an engagement which was broken by
the early death of the young lady.   The acquaintance
of Agassiz with Ampère dated from their constant
meeting at the house of Cuvier.   Next day, late in
the morning — for both Ampère and Agassiz were
confirmed *noctambules*, and they had not retired until
nearly two o'clock in the morning, — their conversation
was resumed; and the whole day passed rapidly, the
only interruption to their reminiscences being a visit
to Mount Auburn Cemetery, where Ampère was much
interested not only in the beauty of the place, but also
in finding the tomb of Spurzheim, an old acquaintance
made in Paris in the twenties; and he lingered also a
long time before the tomb and statue of the mathematician, Nathaniel Bowditch, an old correspondent of
his father.

Postponing his departure from day to day, Ampère,
who came only to pay a visit of a few hours, remained
a week.   Agassiz gave a great dinner party, to which all
the Cambridge professors were invited to meet Ampère,
who thus had an opportunity to see Longfellow, Sparks,

---

[1] Jean Jacques Antoine Ampère, born at Lyon, August 12, 1800, died
at Pau, March 27, 1864.

Felton, Pierce, etc. Longfellow especially pleased him much, and both became absorbed in reciting old French verses, for Ampère was somewhat of a poet himself, and was also professor of old French literature at the Collége de France, at Paris.

Ampère-like, — for his father and he were celebrated for their absent-mindedness, — he stayed eight days, saying every morning that he was going back to Boston; and not only remaining, but, as each day passed, forgetting even to bring a change of linen from his Boston hotel. How he managed to keep up appearances with the same shirt was a problem which furnished great fun to Agassiz's children, who were disposed to see the comical side of their father's extraordinary guest. Every morning Ampère's shirt collar was lowered, in order to conceal the mark of the preceding day, until' the collar wholly disappeared on the last day. Then he found his way back to the Tremont House. Good Ampère! He was a well of knowledge, ready to talk for hours on zoölogy, botany, geology, palæontology, old French, history, political economy, philology, travels, physics, chemistry, poetry, glaciers, fine arts, romance languages (such as Provençal, Italian, Spanish), German literature; in fact, on any subject, like a veritable encyclopædist.

Appointed professor of comparative anatomy at the Medical College of Charleston, South Carolina, Agassiz assumed his new duties in December, 1851, his lectures being delivered every winter, between his autumn and spring courses at Cambridge. He took with him all his family, besides two assistants, Clark and Stimpson, and the artist Burkhardt, — a rather cumbrous establishment so far from Cambridge. Always most hospitably received by the Charleston savants and the "élite" of Charleston society, he found opportunity to deliver not only his courses before the Medical College, three times a week, but also an evening course of lectures to the public in general. But the strain was too great, and his health began to break down. Sullivan's Island, a few miles south of Charleston harbour, where he established his laboratory, did not agree with him. He was constantly feverish, and the South Carolina climate was decidedly unfavourable for him.

Before leaving Charleston he learned with joy that the French Academy of Science of the Institute of France had bestowed on him its first award of the " Prix Cuvier," in consideration of his splendid and difficult work, the " Poissons fossiles." This prize was founded with the money remaining from a public subscription to raise a marble statue of Cuvier in the geological gallery of the Jardin des Plantes, and a monumental fountain at the corner of Cuvier and St. Victor streets, close by the gate of the Jardin des Plantes, in the Pitié Square. It was a reward well bestowed, honouring both the Academy and the recipient.

Agassiz, about this time, had two curious experiences, for which his previous European training had not prepared him. To his great and disagreeable surprise, he found himself entangled in two somewhat serious difficulties almost before he was aware of it. The religious world, always so powerful in America, and more espe-

cially in New England, closely followed Agassiz's teachings. In one of his letters, lately published, Asa Gray
expressed himself as well satisfied with the ground taken
by Agassiz on all spiritual matters; and, thus encouraged by one of his colleagues of Harvard College, who
called himself an "orthodox Presbyterian" of the old
Puritan school, he made no objection to the request of
some of the leaders and editors of the "Christian Examiner" of Boston, to write a few articles for that
periodical; and during 1850 and 1851 he published
three articles on questions of natural history. The
first paper, on "Geographical Distribution of Animals," March, 1850, was well received by every one;
but this was not the case with the next article, "The Diversity of the Origin of the Human Races." Although
he took the scientific ground, and insisted most strongly
that there are two distinct questions involved, — the
Unity of Mankind, and the Diversity of Origin of the
Human Races, — it was taken by some as a sort of support of slavery, and the abolition party became angry
and excited. Finally his third paper, published in January, 1851, "Contemplations of God in the Kosmos,"
seemed to many religious men to make too light of
Genesis, and to pass over Adam as if he had never
existed. Newspapers reviewed the three articles, and
some sharp criticisms were made against Agassiz, not
only in America, but also in Europe. As is often the
case, some, without even reading the articles, took it for
granted that Agassiz wrote them in order to please the
Methodists and the slave-holders; while others, hardly

better informed, accused him of agnosticism, because he mentions " those whose religion consists in a blind adoration of their own construction of the Bible." Having displeased abolitionists, atheists, and pietists, he declined to furnish any more articles to religious periodicals.

His second disagreeable experience at about this same period had to do with savants. The palæontologist of the state of New York came to Cambridge, one day in November, 1849, with a large manuscript " Chart of the Geological Formations," intended for the use of the common schools of the whole state of New York. Agassiz indicated some improvements and additions, and gave a written testimonial. Some time after, he received a copy of another chart of the geological formations made by another person, with a request for his opinion. The sending and request came not directly from the author, but through the palæontologist of New York. Agassiz, accustomed in Europe to give freely his opinions on scientific matters, did not pause an instant to reflect, but wrote a letter disapproving this second chart. Armed with this letter and Agassiz's previous approval of his own chart, the palæontologist of New York succeeded in obtaining from those in authority at Albany the acceptance of his chart[1] and the refusal of the other. The author of the second chart, having learned that the rejection of his chart was due mainly to the opinion expressed by Agassiz

[1] " Key to a Chart of the Successive Geological Formations, with an Actual Section from the Atlantic to the Pacific Ocean. Illustrated by the Characteristic Fossils of Each Formation." By James Hall, Boston, 1852.

in a private letter, sued Agassiz for damages before the court of justice at Albany. A rather long and expensive trial followed; and although Agassiz won his case, and showed that a savant had a right to give his opinion upon any published scientific subject, he was considerably annoyed by the proceedings, and never afterward recommended anything for publication. The truth is, that both charts were poor, and the choice between the two was rather embarrassing on that account. Agassiz's good faith, reputation, and friendship have been too often used for money-making without scruple, and not always for the good of science.

CHAPTER XV.

1852 (*continued*)–1855.

AFTER leaving Charleston, Agassiz stopped at Washington, to deliver a course of lectures at the Smithsonian Institution, on the "Foundation of Symmetry in the Animal Kingdom." He had already lectured once before, at the Smithsonian, in 1850, on the "Unity of the Plan of the Animal Kingdom." The great Institution, which has since done so much for American science, and indeed for the whole world, was then in its infancy, for it had begun its operations soon after Agassiz's arrival in America, the "First Report" of the secretary, Joseph Henry, being dated Dec. 8, 1847. Professor Henry was much attracted by Agassiz's immense store of scientific knowledge and his great experience of European academies, scientific societies, scientific journals, and personal acquaintance with all the leaders, — from George Cuvier, to Humboldt, Arago, and Leverrier, — and quickly took advantage of his presence in America to become intimate with him, after a few years of acquaintance. The apartment he occupied in the Smithsonian Building was thrown open to Agassiz, as often as he visited or passed through Washington. Agassiz enjoyed in Henry not only his great capacity as an original observer in experimental physics and meteorology, but also his capability as a scientific administrator. Everything was to be done at once; and Henry was very glad to be able to consult Agassiz on everything connected with natural history, great collections, scientific libraries, and relations with foreign societies, institutions, and scientific bodies. The first list of foreign academies and scientific societies was suggested in a great measure by Agassiz, who asked me to help him. The instructions for collecting natural history subjects were partly translated from the "Instructions pour les voyageurs," par l'administration du Muséum royal d'Histoire naturelle ($4^{i\text{ème}}$ édition, Paris, 1845), of which I gladly offered copies to Agassiz and Henry. Of course, Agassiz, Baird, and Girard added a great deal to those instructions, which have contributed so much toward making the United States National Museum the greatest depository of American natural history specimens.

Although constantly in relation with Henry and his assistant, Baird, Agassiz was not appointed one of the regents of the Smithsonian Institution until Feb. 6, 1863, in place of Mr. Badger, removed, as a "traitor," during the Civil War. In the Annual Report of 1862, printed in 1863, Agassiz is for the first time on the list of regents, being the last of the list; and from that time until his death he was a constant attendant at the meetings, taking great interest in all that related to the Institution.

Professor Alexander Dallas Bache, the justly celebrated director of the United States Coast Survey, was the first American savant to appreciate what a valuable addition Agassiz was to American science; and he at once put at his disposal all the vessels and steamers employed in surveying the Atlantic coast. A very strong friendship rapidly sprang up between them, and though the two men were entirely different, they admirably supplemented one another. Bache was a good and accurate mathematician, and inherited from his grandfather, Benjamin Franklin, great administrative power, — two things entirely wanting in Agassiz, who knew absolutely nothing of mathematics, or even of arithmetic, and was a rather poor administrator, as we have seen.

Bache preceded Henry at Washington by three years, having been appointed Professor Hassler's successor as superintendent of the United States Coast Survey, in December, 1843. Under his direction, the bureau became very important; and he had the good judgment to choose for his principal assistants the most able young officers of the army, such as Major Isaac I. Stevens, afterwards governor of Washington Territory, and major-general, United States Volunteers; Lieutenant A. A. Humphrey, afterward chief of the staff of Meade during the last Virginia campaign, and brigadier-general and chief of engineers,

United States army; and many others. The publications of the Coast Survey, which, until then, were
limited to marine charts, were largely extended to
geographical subjects, astronomical expeditions, and
studies of the coal reefs, and tidal researches.

For many years, we may say that the triumvirate of
Bache, Henry, and Agassiz led American science, and,
on the whole, they gave the strongest impulse science
has received on this side of the Atlantic.

Life in Cambridge was resumed with great relish by
Agassiz. He was full of schemes for new researches and
publications, and his activity was as great as it had ever
been before. Society claimed his presence in Cambridge
and in Boston, and, as he was very fond of social occasions, he accepted all invitations. He became very
popular with the members of the various clubs which
he joined, and his presence enlivened the tables of all
the "élite." Felton and he were inseparable, and it was
a pleasure to hear them in after-dinner talk. Agassiz
was very genial and would talk for hours; Felton was
also full of anecdote; and both were charming companions. They had royal times together, rarely returning home until one or two o'clock in the morning.
Such late hours made early rising out of the question,
and Agassiz was seldom at his breakfast table before
eleven o'clock, often not before twelve o'clock. Then,
after lighting a cigar, he would start for his laboratory,
where he would examine some wonderful organisms
with the microscope, directing the attention of his
pupils to some special point, correct their drawings,
and encourage them in every way; for he had no equal

in the art of instigating researches, and inspiring his hearers with desires to accomplish something grand and new to science.

The pupils of Agassiz in America may be divided into two series : the first dates from his arrival until the opening of his Museum of Comparative Zoölogy in 1860, and the second, from the opening of the Museum until his death. I shall only notice those who have gained celebrity in the scientific world.

His first pupil was his son Alexander, upon whom Agassiz bestowed much of his time. Of course, in this case, paternal love was interested, and some anxiety was caused, when, after graduation from college, and lecturing for two years at his school, Alexander entered the United States Coast Survey as an assistant, and departed for a survey of the mouth of the Columbia River (Oregon). However, he returned in July, 1860, a few months before the opening of the Zoölogical Museum, and devoted himself to the work of arranging the collections of animals preserved in alcohol, — by no means an easy task. His success in studying marine animals, and more especially echinoderms, was a great pleasure to his father, who was justly proud of his beautiful and excellent monograph, " Revision of the Echinoderms," Cambridge, 1872. Since this time Mr. Alexander Agassiz has become an expert and an authority on animals obtained from deep-sea soundings, not only in the Gulf of Mexico and along the Atlantic coast, but also in the Pacific Ocean ; and he is considered the best specialist on living echinoderms.

The second pupil of this first set was William Stimpson, an extremely bright young Cambridge student, with no small amount of originality. Stimpson took at once to dredging along the sea-bottom, in order to investigate and determine provinces of marine life, more especially of the Mollusca. He was strongly impressed by Edward Forbes's researches in that line, and followed in his steps, not only on the shores of the British Isles, but on the coasts of Maine, Massachusetts, Maryland, Virginia, the Carolinas, and Florida, adding greatly to that branch of natural history. After leaving Professor Agassiz's laboratory in 1855, Stimpson was attached to the Smithsonian Institution at Washington, and did much, in collaboration with Dr. Charles Girard, to create the collection of living marine invertebrates of the United States National Museum. Having agreed to found and direct a museum of natural history at Chicago, he had the misfortune to see all his manuscripts and collections destroyed by the great Chicago fire of October, 1871. Although Stimpson died young, he left an imperishable name in conchology.

H. James Clark, also a graduate like Stimpson of the Lawrence Scientific School of Harvard University, was the favourite pupil of Agassiz; his investigation of the embryology of turtles, and his microscopic illustrations of all the researches of Agassiz, contained in his "Contributions to the Natural History of the United States of America," show a rare amount of patience, and great accuracy as an original observer. Clark possessed a quality which was much admired by Agassiz; namely, steadiness in work. He was indefatigable at the microscope, day after day, month after month, year after

year. In the eyes of Agassiz, everything and every
one in his laboratory was second to Mr. Clark. In the
construction of his house on Quincy Street, in 1855, he
took special care to have a stone pillar placed where it
would receive the best northern light for Clark's microscope. He did the same when, in 1860, he built his
great museum. In fact, Clark was his right hand during almost twelve years. I quote a letter from Agassiz
which will show the great place Clark occupied in the
scientific organization of Agassiz's establishment : —

CAMBRIDGE, 24 juillet, 1860.

MON CHER PICTET (Jules Pictet de la Rive, à Genève).

C'est un vrai plaisir pour moi de vous présenter mon collègue, Mr.
H. J. Clark, celui de tous mes élèves dont j'attends le plus. Vous
verrez bien qu'il a embrassé l'histoire naturelle dans son ensemble,
et je ne crois pas qu'il existe un naturaliste, plus habile que lui dans
l'emploi du microscope.

Tout à vous,

LS. AGASSIZ.

Clark was appointed adjunct professor of zoölogy at
Harvard University, on the special recommendation of
Agassiz. As had happened before, Agassiz, with his
enthusiastic and sanguine temperament, had raised
hopes of pecuniary position in Clark's mind, as soon
as his great museum should be inaugurated, which it
was impossible to gratify, at least immediately. Disappointed in his expectations, and with a large family to
provide for, Clark's conduct was such that he was
obliged to resign his position at Harvard ; the difficulty
having become so personal that Agassiz simply said to
the Board of Trustees that he or Clark must leave. In
such a dilemma, the question was of necessity decided

against Clark.   The case was a particularly trying one,
and is much to be regretted.   Poorly paid, — receiving
hardly enough to sustain his family, — Clark thought
that the least he could receive from Agassiz was public acknowledgment of collaboration on the title-pages
of the second, third, and fourth volumes of the "Contributions to the Natural History of the United States."
Instead of this, Agassiz contented himself with saying
in the prefaces that he had "received much valuable assistance" from his friend and colleague, "Professor H. J. Clark," who had "assisted him from the
beginning of his investigations of the embryology of
these animals with untiring patience and unsurpassed
accuracy."   On the point of authorship, Agassiz was
very sensitive and easily offended, and would not allow
any one to interfere.   Clark asked him in vain to refer
his claim to "an arbitration by competent umpires."
Agassiz declined to consent to such a demand, which
he considered as rather preposterous from an old
pupil.

Then Clark published in July, 1863, a small pamphlet
of three pages only : "A Claim for Scientific Property,"
over his signature and title of adjunct professor of Harvard University, which brought the affair to a crisis and
caused his dismissal.   Clark never rallied from the
shock, and died at Amherst the 1st of July, 1873, a
few months before Agassiz's death.   I do not hesitate
to say that this was the most unfortunate scientific difficulty with which Agassiz was connected.   Although he
was right in the main, he might have shown more len-

iency to a favourite pupil, and have given him full satisfaction, by inscribing his name as collaborator on the
title-page of the three volumes, without diminishing in
any way his own share in the work. Cuvier did so with
Valenciennes, in his " Histoire naturelle des Poissons";
and with that precedent Agassiz might have granted
Clark's claim.

The case of Karl Vogt's claim in regard to the
"Anatomie des Salmones" is different; for Vogt had
his name recorded as collaborator on the title of the
work. Had Agassiz done the same with Clark, he
would have raised himself above the petty question of
scientific ownership of a few observations or thoughts,
which it is always very difficult to decide in the case of
two observers engaged together on the same work and
daily exchanging their views.

Mr. James E. Mills, who worked out for Agassiz the
special characters of the families of turtles, removed to
California in 1858, where he has since lived, engaged
in gold mining and practical geological work in the
Sierra Nevada and other parts of Central California, as
well as in Brazil.

Dr. David F. Weinland, a German from Frankfort-on-
the-Main, helped Agassiz in his work on the "Anatomy of the Turtles," between 1856–58, and is referred
to in the Preface, p. xv, of "Contributions to the Natural History of the United States," Vol. I.

Theodore Lyman, who was graduated in the same
year as Alexander Agassiz, 1856, was another favourite
pupil of Agassiz. In 1857, he made an exploration of

Florida, returning with a rich collection of echinoderms, corals, etc., which he presented to the Museum.
He has devoted most of his zoölogical researches to
the interesting and beautiful families of the Asteroides
and the Crinoides of the present fauna, though during
the last twenty-five years the number of Crinoides has
increased to such an extent that he has been hardly
able to describe the new species brought up from the
deep seas by the different expeditions. Lyman is the
authority for everything relating to living Crinoides;
but unhappily, on account of his health, he has been
obliged to give up all work at an age when it was expected that he would much increase our knowledge as
well as assist Congress, of which he was a member for
Massachusetts, in the reform, so much needed, of the
scientific organizations of the United States government.

The following letter from Agassiz to his friend,
Jules Pictet de la Rive, of Geneva, will give an idea
of the high esteem and friendship he felt for Theodore
Lyman : —

CAMBRIDGE, 11 juin, 1861.

Mon cher ami, — Permettez que je vous présente Mr. Th.
Lyman de Boston, un de mes élèves de prédilection, et beau-frère
de mon fils, qui se rend en Europe avec sa femme pour voir le
monde, et faire la connaissance des savants d'outre-mer. Je vous
le recommande tout particulièrement comme mon ami et comme un
géologiste plein d'avenir.

Votre tout dévoué,

L3. ΛΟΛCΕΙΖ.

Fred. W. Putnam, of Salem, joined the laboratory

and class of Agassiz at the beginning of 1856, making a specialty of living fishes, and took charge of that important branch of the collections when the Agassiz Museum was inaugurated in 1860. He has since entirely given up his zoölogical work, and turned to ethnology and prehistoric man, and is now director of the Peabody Museum at Harvard University.

I may add to this first list of Agassiz's pupils in America the entomologist, Dr. John L. Leconte, who often came to Cambridge as a guest of Professor Agassiz, and learned much from the great store of knowledge and methods of studying of the professor; and as an acknowledgment of the benefits received from these visits, he bestowed his large, important, and rich entomological collection upon the Agassiz Museum. Dr. Joseph Leconte, a cousin, was also a pupil at Charleston, South Carolina, and also at Cambridge. He has since published a Manual of geology, based in part on the lectures of Professor Agassiz, which has given him a certain reputation, and he has also become professor of geology at the State University of California.

Agassiz's annual visit to Charleston, South Carolina, was attended by most serious illness, — an attack of that southern fever generally called *malaria*, — which brought him to death's door. His life was at moments despaired of, and he was in great danger for many days. This illness incapacitated him for two months, from Christmas, 1852, until the end of February, 1853, when, with his unconquerable energy, he again began to deliver his lectures before the Medical

School. The Charleston climate had always disagreed with him; and at each of his four visits, from 1847 to 1852, he suffered an attack of sickness, either while there or as soon as he left. His South Carolina friends, Drs. Holbrook and Ravenel, who had taken such good care of him during his last illness, advised him not to return, and he consequently resigned his professorship at the Medical College.

As soon as his lectures were finished, he started for a prolonged tour in the South, delivering lectures at Mobile, New Orleans, and St. Louis. The Mississippi River was a wonder to him, with its muddy waters, and its rich fauna of fishes, turtles, and caimans. He had there, on a smaller scale, the spectacle which so much impressed Spix on the Amazon, and which had haunted him ever since he had described Spix's fishes at Munich in 1829. The journey up the Mississippi increased, if possible, his desire to explore the Amazon; a desire which he finally realized fourteen years later.

By this time his house on Oxford Street was over-crowded by inhabitants, books, and all sorts of *imped-imenta*. Agassiz was still so full of future work, and he was so eager to accumulate materials of all sorts for studies, that he brought home everything he could lay his hands upon. As an illustration I may give a per-sonal recollection. During May, 1853, he drove to my house in Dorchester, and packed his two-horse carriage full of my books, such as the publications of the Geo-logical Society of London, a full set of the reports of the British Association for the Advancement of Science,

and a quantity of volumes on the geology and palæontology of France and Italy.   As I was on the eve of
an exploration from the Mississippi River to the Pacific
Ocean for the United States government, and expected
to be absent at least a year, Agassiz thought that he
might want to consult many of my books during my
long absence, and he therefore carried them to his
house.   This shows also how scarce scientific books
were then in America.   The few savants scattered
over New England were obliged to borrow from one
another the memoirs they wished to consult in their
work.   Now with such rich libraries as we have at our
disposal, it seems hardly possible that only forty years
separate us from that time of difficulty in consulting all
the publications needed for a special study.   Agassiz
was thinking of those times when, eight years later,
after he had gathered a valuable natural history library
at his museum, he generously offered to allow American
naturalists to borrow all the books they wanted.

A larger house had become an absolute necessity;
and accordingly Harvard College built one for him, on
a piece of its ground at the corner of Quincy and Harvard streets, just opposite the house of his friends, the
Feltons.   He left the second Hôtel des Neuchâtelois
in Oxford Street during 1854, after a sojourn of seven
years; and it may be said that Agassiz passed there
the happiest time of his life.   For there he was freed
from that sort of nightmare which hung over him so
long, his abnormal and never well-defined association
with his secretary, Desor.   There he received and lived

with his children; while living there he was married, and there also he entertained all his American and European friends.

The main difficulty that Mrs. Agassiz had to contend with as soon as she entered the Oxford Street house was to obtain a regular supply of money for daily household expenses.   At last she realized that it was almost hopeless to expect a reform of Agassiz in this direction, and she herself took the matter in hand. · With the help of the two oldest children, Alexander and Ida, she decided to open a school for young ladies and girls, and to locate it in the upper story of the great house in Quincy Street.   Agassiz, whose strongest passion had always been for teaching, was enchanted with the scheme, and entered into it with great enthusiasm.

Mrs. Agassiz had the whole management of the school; everything was referred to her, as director. It is important to remark that she had had absolutely no experience in teaching, either in a public or private school.   She had received her education from an English governess in her family, and did not enjoy the advantage of a school education bestowed upon almost all American girls.   Nevertheless, she took the directorship of Agassiz's school in a masterly way, and succeeded admirably.   She herself did not teach, but everything regarding the teaching came under her supervision. As the fees were high, the school was a very select one; and pupils came from different parts of the United States, even from as far west as St. Louis.   It was considered a great privilege to be taught by such a naturalist as Agassiz, and all the girls whose parents could afford it were anxious to join the school.

Of course the great attraction was Agassiz, who lectured every day of the week, except Saturday. The girls' parents often came with them, and sat down in the schoolroom to listen to the lectures, which were so clear and so entertaining, that every one followed, with the greatest attention, the subjects brought up by their great teacher, however difficult they might be. But it must be said that, although the school continued eight years, and the number of pupils who passed through it was quite large, — about five hundred, — not a single one of them became a naturalist, or even an "amateur" in natural history. The only female pupil Agassiz made in all his life was his second wife; and even she gave up her studies in this line after his death, showing that it was not through inclination and special taste that she had become a naturalist, but only through her husband's inspiration.

The money brought in by the school was a great help and a great relief. As Mrs. Agassiz says, "He was never again involved in the pecuniary anxieties of his earlier career." [1] However, it must not be supposed that from the day he opened his school for girls he had no further money difficulties. It was impossible for a man of his nature to keep free from such difficulties. As long as he lived, he was constantly hunting after a dollar to pay some expense he had already incurred.

[1] "Louis Agassiz," by Mrs. E. C. Agassiz, Vol. II., p. 527.

On the whole, the school for girls was a most successful undertaking, which reflected great credit on
the leadership of Mrs. Agassiz, and on the practical
turn of mind shown by Alexander Agassiz and Miss
Ida.

CHAPTER XVI.

1856–1858.

UNDER the direction of Mr. Francis C. Gray, of Boston, a most generous and constant friend of Agassiz, the editorship of an important and costly work on the natural history of North America was undertaken. A subscription list was started as soon as the prospectus was issued, in the autumn of 1855, and, to the astonishment and great delight of Agassiz, quickly reached the unexpected number of twenty-five hundred subscribers, the necessary number suggested by the publisher, to insure the success of the publication, being five hundred. The subscription price was twelve dollars per volume, and there were to be ten volumes, each volume being entirely independent, except the first two, which were combined in such a manner that they formed a whole. The first two volumes were issued in April, 1858; two more volumes appeared in

1860 and 1862; and then the work was interrupted, and never resumed.

The large number of subscribers to such a costly and special publication proves the great popularity attained by Agassiz during the first eight years of his stay in America. He had succeeded in exciting an interest in questions of natural history, until then much neglected, not only as a special pursuit, but as a part of the general education of the people at large. The citizens liberally showed their interest in the undertaking, not only because they thought that the subject was worthy, but also to reward a naturalist of world renown, whom they wanted to attach more and more to themselves, and persuade to make America his home and adopted country. This explains both the success of the subscription, and the following optimistic sentences in the preface of Volume I.: "I must beg my European readers to remember that this work is written in America, and more especially for America; and that the community to which it is particularly addressed has very different wants from those of the reading public in Europe. There is not a class of learned men here, distinct from the other cultivated members of the community. On the contrary, so general is the desire for knowledge, that I expect to see my book read by operatives, by fishermen, by farmers, quite as extensively as by the students in our colleges, or by the learned professions; and it is but proper that I should endeavour to make myself understood by all" ("Contributions Natural History of the United States," Vol. I., Preface, p. x).

His expectation proved a Utopian dream; for, except
Part I., "Essay on Classification," all the other memoirs,
— on the "Turtles," on the "Acalephs," on the "Radiata," — are so special, that only very few persons were
able to read them with anything approaching a general
understanding, and fewer still, to follow the minute
descriptions. It is no exaggeration to say that the
number of persons in America who read this great
work, except the "Essay on Classification," was limited
to less than one hundred; and that the specialists in
Europe interested in the subjects treated numbered
only a few dozens.   However, it should be said that
these memoirs are worthy of Agassiz's great reputation
as a naturalist, and have added many new facts in
regard to the Testudinata and Acalephs.   The figures
on the plates are all excellent, and show beyond question that natural history specimens and details of the
most delicate anatomical structures were treated in a
style which was never surpassed and rarely equalled in
Europe.

Part I., "Essay on Classification," was read by many.
To say that it was understood, in all its meaning and
far-reaching generalities, would be wide of the mark.
It requires a profound and vast knowledge of natural
history enjoyed by few naturalists to understand such
a philosophical work, which is in fact a *résumé* of the
discoveries of all observers since Linnæus and Cuvier.

A special octavo edition of the "Essay" was reprinted in England, and a French translation in Paris.
It is the work of the mature age of Cuvier's best pupil,
and is by far the most important contribution of Agas-

siz to natural history during his life in America. It
contains the last great discovery he made; namely,
that "the changes which animals undergo during their
embryonic growth coincide with the order of succession
of the fossils of the same type in past geological ages."

If the influence exerted by the "Essay" was not so
great as it should have been, it was due to adverse
circumstances which it was impossible to foresee and
prevent. During his stay in Europe, Agassiz's researches
were mainly palæontologic, with the study of glaciers
as a sort of recreation. His true zoölogical studies were
confined to fresh-water fishes, and even these studies he
did not carry very far. In America, on the contrary,
he devoted almost all his time to zoölogy and to embryologic researches, almost entirely abandoning palæontology and glaciers; and it required ten years of hard and
continuous work, mainly with the microscope, to enable
him to master the purely zoölogical part of living
animals, and explain its harmony with the palæontology.
It is not just to reproach him, as has sometimes been
done, with the fact that his work on classification came
too late by ten years; for it was impossible for him to
collect sooner the immense quantity of materials required, though, of course, if it had been published ten
years earlier, it would have exerted a greater influence
on his contemporaries.

On the other hand, the publication of Darwin's
"Origin of Species," only two years after the issue of
Agassiz's "Essay on Classification," distracted the attention of a certain number of savants, who seized this
opportunity to discuss and checkmate the theory of

diversity of creations, admitted and propagated by Cuvier's school. The great misfortune of Agassiz's " Essay " was that it came at an inopportune moment. It was too late to do all the good that it ought to have done ; and too soon, because of the discussion and passionate polemic raised by Darwin. If Agassiz had waited three years longer, he would have given another shape to his great generalization, and presented well-digested views in opposition to the Lamarckian system revived by Darwin. It is to be regretted that Agassiz entered into personal encounters at meetings of academies and scientific societies, for, not being a good debater like Cuvier, he failed to present the best part of his argument; whereas, if in the calm of his library, he had marshalled all his facts against natural selection, the survival of the fittest, etc., he would have exerted a very beneficial influence in the sort of unreasonable allurement which induced a large number of semi-savants to enter the path reopened by Darwin, to the conquest of the creation and the Creator.

As it is, however, Agassiz's " Essay" is a great work, and will remain in the history of classification. Pecuniarially, also, it was a great success for both Agassiz and the publishers.

The issue of the first two volumes almost coincide with the anniversary of his fiftieth birthday. At least on that day, the 27th of May, 1857, the manuscript was so far advanced, that he might justly have felt that he had attained the end of his task. While hard at work, at his library desk, as the clock struck twelve, musicians stationed in front of his house began a serenade which was followed by congratulations of friends and special students. It was for this occasion that his colleague and friend, Longfellow, composed the following verses: —

THE FIFTIETH BIRTHDAY OF AGASSIZ.

It was fifty years ago,
   In the pleasant month of May,
In the beautiful Pays de Vaud,
   A child in its cradle lay.

And Nature, the old nurse, took
   The child upon her knee,
Saying: "Here is a story-book
   Thy Father has written for thee.

"Come wander with me," she said,
   "Into regions yet untrod;
And read what is still unread
   In the manuscripts of God."

And he wandered away and away
   With Nature, the dear old nurse,
Who sang to him night and day
   The rhymes of the universe.

And whenever the way seemed long,
   Or his heart began to fail,
She would sing a more wonderful song,
   Or tell a more marvellous tale.

So she keeps him still a child,
   And will not let him go,
Though at times his heart beats wild
   For the beautiful Pays de Vaud, —

Though at times he hears in his dreams
   The Ranz des Vaches of old,
And the rush of mountain streams
   From glaciers clear and cold;

> And the mother at home says, " Hark!
> For his voice I listen and yearn ;
> It is growing late and dark,
> And my boy does not return !"

May 28, 1857.

When Switzerland founded a federal polytechnic
school, as a sort of compromise for a state university,
Agassiz was written to, unofficially, in regard to an
appointment. The friend who sent the message, the
learned Oswald Heer, wrote first to call Agassiz's attention to the advantages to be derived from a position
among his old friends, some of them his classmates also,
as Arnold Escher de la Linth, Albert Mousson, and his
first teacher in zoölogy, the old Schink, who was still
living at Zürich ; and second, at the same time, to offer
to sell Agassiz his own private collection of Oeningen
fossils, knowing well how easily he was tempted by
collections. But Agassiz, in a letter dated January,
1855, declined both offers, at the same time asking for
Glaris fossils fishes, if Heer was able to procure any,
and saying that as soon as he had money at his command, he would with pleasure purchase his collection of
Oeningen fossils, which he was enabled to do five years
later.

But a much more tempting offer was made in August,
1857, when Agassiz received the following official letter :

PARIS, le 19 août, 1857.

Monsieur, — Une chaire de paléontologie est vacante au Muséum
d'Histoire Naturelle de Paris.   Vous êtes Français, vous avez en-
richi votre pays natal de travaux éminents et de recherches labo-
rieuses ; vous êtes membre correspondant de l'Institut. L'Empereur
serait heureux de ramener en France un savant distingué, un pro-

fesseur **renommé**. Je viens vous offrir en son nom la chaire vacante
et votre patrie se félicitera de retrouver un de ses enfants les plus
dévoués à la science.

Veuillez agréer, Monsieur, l'assurance de mes sentiments de haute
estime.

ROULAND.

The Museum of Natural History or Jardin des Plantes
had just passed through a great crisis. Its organization
needed a complete reform; but two successive commit-
tees, appointed in 1849 and in 1858, to report on the
condition and improvements to be introduced, in order
to end the existing anarchy, had most pitifully failed to
do anything, owing to the factious opposition of the
professors, who were at the same time administrators of
that great establishment. The Emperor and his ministry
were well acquainted with the difficulties, owing to information obtained from a naturalist of talent, Prince of
Canino, Charles Lucien Bonaparte, who was mentioned
by Napoleon III. for the directorship, with power to
form a new organization and to put into operation all
the needed reforms to keep the establishment on a level
with foreign institutions of the same sort, or even at
their head, as it had been during the time of Cuvier,
Lamarck, Geoffroy Saint-Hilaire, Lacépède, Desfon-
taines, etc.

Prince Charles Lucien Bonaparte had spoken to his
cousin, the Emperor, of the great value of Agassiz; he
had always maintained intercourse with him, ever since
they had proposed a joint expedition to the United
States in 1842, and he relied much on Agassiz to reform
the Jardin des Plantes; but his death in 1857, at the
premature age of fifty-four, put an end to the scheme.

The French government, however, resolved to adopt his views of reform, and the Secretary of Public Instruction wrote the preceding letter, hoping that Agassiz would accept and accomplish the complete reorganization so much needed in the Museum of Natural History.

But it was too late; the French government had twice missed its opportunity. The offer should have been made in 1846, when Agassiz was in Paris, poor and anxious as to his future; or even in 1853, on his return from Charleston, after the serious illness which endangered his life. Now, after the great success of his school for girls, of the subscription to his great work on the natural history of North America, with brilliant prospects for the foundation of a great museum, which should be entirely his own, and with the strong family ties resulting from his second marriage, it was out of the question for him to return to Europe and begin life a third time, however attractive the offer and prospect might be.

After long and deliberate consideration, Agassiz declined, in the following letter to M. Rouland: —

À Son Excellence le Ministre de l'Instruction Publique et des Cultes, à Paris.

Monsieur le Ministre, — Après avoir passé la plus grande partie de ma vie éloigné des grands centres scientifiques, je ne me serais jamais attendu à recevoir l'honneur très distingué que vous m'avez fait, en m'offrant au nom de l'Empereur, la chaire de Paléontologie au Muséum d'Histoire Naturelle de Paris.

Le monde entier regarde le Jardin des Plantes comme l'établissement le plus important qui existe pour les sciences naturelles. Aussi ai-je lu votre lettre avec le plus grand plaisir, et en recevant votre offre j'ai eu la preuve, bien précieuse pour moi, que je n'étais

pas oublié en Europe. Malheureusement je me trouve dans l'incapacité d'accepter votre proposition, car il m'est impossible de trancher brusquement les liens que depuis plusieurs années je me suis
habitué à considérer comme m'attachant aux États-Unis pour le
reste de mes jours. Comme je ne puis supposer que l'enseignement
qui était confié à M. d'Orbigny puisse être interrompu assez longtemps pour me permettre de finir les travaux embryologiques que
j'ai entrepris en vue d'établir des comparaisons avec les fossiles des
époques antérieures à la nôtre, travaux qui perdraient tout leur
intérêt si je les laissais inachevés, je me trouve ainsi placé dans la
pénible nécessité de refuser une position que, dans toutes les circonstances, je regarderai toujours comme la plus brillante à laquelle
un naturaliste puisse aspirer.

Il peut vous paraître étrange que je laisse quelques ovaires et
embryons peser dans la balance qui doit décider du reste de ma vie ;
mais c'est sans aucun doute à ce dévouement absolu à l'étude de la
nature que je dois la confiance dont vous venez de me donner une
marque aussi éclatante qu'inattendue ; et c'est précisément parce
que je désire de continuer à la mériter dans l'avenir, que j'ai pris la
liberté d'entrer dans ces détails.

Permettez-moi aussi de rectifier une erreur qui circule sur moi.
Je ne suis pas Français. Quoique d'origine française, ma famille a
été Suisse depuis des siècles ; et moi-même, malgré une expatriation
qui dure depuis plus de dix années, je n'ai pas cessé d'être Suisse.

Je demande à Votre Excellence de recevoir, avec le renouvellement de mes regrets sincères de ne pouvoir accepter la chaire que
vous m'offrez, l'assurance de ma considération la plus distinguée.

LOUIS AGASSIZ,
Professeur à l'Université de Cambridge
(*États-Unis d'Amérique*).
CAMBRIDGE, le 25 septembre, 1857.

M. Rouland was so desirous of securing Agassiz's
services that he did not accept this refusal as final, but
wrote him again, saying that he would let his offer
stand for two years, in order to allow him to finish his

most important works in America; and it was not
until Agassiz's personal visit to him at Paris, in July,
1859, during which the Secretary of Public Instructions
explained to him that the chair of palæontology was
only a pretext to bring him to France, that he accepted
his decision as final. He intimated to him that he
would have the directorship of the Museum of Natural
History,—an office to be created,—very likely another
chair at the Collége de France, as Cuvier had, and
finally a senatorship, with salaries amounting to not far
from fifteen thousand dollars, —an offer brilliant, both
as to rank and remuneration.

But it was all in vain. Agassiz's refusal was final. The
French government behaved nobly, taking the refusal
in good part, and continuing to consult him on questions of acclimatization of marine animals, and sending
him, in succession, the cross of a Knight of the Legion
of Honour and of an officer of the same order.   In brief,
it may be said that Agassiz received from France more
tangible proofs of the great esteem in which he was
held than from any other European country. He was
first elected correspondent,[1] afterward foreign associate
fellow (Membre associé étranger) of the Academy of
Science of the Institute, a rare and much-valued distinction. He received a Monthyon prize of Physiology and
the Cuvier prize from the same academy, was offered
officially the chair of palæontology at the Jardin des

[1] Agassiz was elected a corresponding member of the Academy of Science of France as far back as April, 1839, when he was barely thirty-two
years old. His concurrent was Prince Canino, the son of Lucien Bonaparte,
and the election was very close.

Plantes, and was created Knight and officer of the Legion of Honour. In 1870–1871, when France was in trouble and suffering such crushing defeats, although mainly through her own fault, Agassiz came forward, and expressed his great disapproval of the brutal con-duct of the victorious Prussians, fully realizing the debt the present civilization and political freedom in Europe owed to the many good acts and the intervention of France in behalf of progress. Indeed, without the help offered so generously to Switzerland, in 1857, his own canton of Neuchâtel, of which he was a burgess (Bourgeois de Neuchâtel et de Valengin), would have remained to this day under the rule of the king of Prussia. Although Agassiz shrank all his life from politics, he was very liberal, and always in favour of liberty.

One of Agassiz's first schemes when he came to the New World was the preparation of a great work on the fresh-water fishes, analogous to the one he had attempted in Europe on the same subject. He soon discovered that the best man to associate with him to carry out his inten-tions was Dr. Spencer F. Baird, then a young professor at Carlisle, Pennsylvania. Baird entered with great enthu-siasm into all the views and ideas of Agassiz, explained to him during a prolonged visit to Cambridge in May and June, 1848; and after an understanding in regard to the number of specimens to be collected and their geographical distribution, he started to explore Lake Champlain. Baird succeeded, but he did one thing which he had not anticipated. He made the acquaint-ance, at Burlington, of Senator George P. Marsh of Vermont, one of the regents and a member of the executive committee of the Smithsonian Institution, who was so impressed by the knowledge, modest bearing, and industry of the young ichthyologist that he proposed him as assistant director of the new institution. After some years of dilatoriness and continuous postponement, really due to want of time, Baird saw that it was idle to expect the realization of Agassiz's scheme, and abandoned it altogether. Agassiz, however, was too much of a fish-lover not to be constantly reminded that a great work was in reserve for him; and during his journey up the Mississippi from New Orleans to St. Louis in 1853, he was much struck by the differences of fishes, according to difference of latitude, in that long north-south watercourse, and in accordance with a habit formed at that time, and practised constantly afterward, he printed a circular asking for information and for collections, which was distributed largely all along the courses of the rivers and on the coasts of the Lakes. Answers came by dozens, and collections followed one another, until Agassiz had ichthyological treasures to his heart's content. But time to co-ordinate and make use of all these facts and specimens was lacking, and one scheme after another constantly postponed the promised study of the distribution and localization of the fresh-water fishes of the United States. Only a few short papers were published on "Extraordinary Fishes from California," on "Fishes from the Southern Bend of the Tennessee River, Alabama," while some of his letters on the subject have appeared in Mrs. Agassiz's life of her husband; but this is all.

CHAPTER XVII.

1858–1864.

IN June, 1859, Agassiz, in company with his wife
and youngest daughter, left for Europe. He wished
to see his old mother, and to present his American
wife to her and to all the members of his Swiss and
German families. On his way to Switzerland, he lingered a few days in the British Isles to see his old
friends, the Earl of Enniskillen and Sir Philip Egerton, the two most distinguished palæoichthyologists of
Great Britain and Ireland; Richard Owen, at Richmond Park, near London; and Roderick Murchison,
who invited all the naturalists then in London to meet
him at his fine house in the fashionable and aristocratic
West End.

At Paris a week quickly passed among Agassiz's old
friends, the French savants; and he had long conver-
sations with the Secretary of Public Instruction, M.
Rouland, on all sorts of questions relating to natural

history establishments and organizations. Although most cordially received by the French naturalists, Agassiz easily perceived the presence of a fear that he would accept M. Rouland's previous offers ; but he foresaw too many difficulties to yield even to such a tempting proposal. As he said afterward, it would have been impossible to reform anything at the Jardin des Plantes, without deeply and irremediably wounding his friends, Valenciennes, Henry Milne-Edwards, Decaisne, and others. When they knew his final refusal, it was a great relief to them all, and their pleasure in his visit was increased a hundred fold.

His return to Switzerland gave immense pleasure to Agassiz, who, like a true Swiss boy, shed tears when he again beheld the Alps. His distinguished mother was enraptured to see her favourite child again, and the few weeks they spent together passed like a dream for her. Agassiz spent most of his time in her company at the beautiful country house of her eldest daughter, Mrs. Cécile Wagnon, at Montagny, between Yverdun and Grandson, "au beau Pays de Vaud"; and, on seeing them together, it was most evident, as one of the family told me a short time afterward, that the great attractiveness possessed by Agassiz was a gift from his mother. They spent most of their time in the "buvette," a sort of out-of-door sitting-room in the garden ; and there, with the great panorama of the Alps in the distance, and back of him the Aiguilles de Baulmes and the Suchet, two of the old Jura mountains where he used to hunt for plants, boulders, and traces of old glaciers, Agassiz passed a summer of repose and true happiness.   He visited at Lausanne his younger sister, Mrs. Olympe Francillon, and other family relations.

The Italian war, then raging, had much disturbed all Switzerland.   The Helvetic Society of Natural Sciences, which met every year in some part of the Swiss Confederation, was to have had its session at Lugano, in Tessino, close by the boundary line of Italy.   On account of the war, the meeting was postponed to the next year; but the arrival of Agassiz, and also the sudden conclusion of peace between France and Austria, gave another turn to affairs.   It was too late for a regular call of the Helvetic Society; but Pictet had the happy thought of asking the " Société de Physique et d'Histoire Naturelle de Genève," to take upon itself the calling of an extraordinary meeting of the Helvetic Society of the Natural Sciences, in order to allow the numerous scientific friends of Agassiz, scattered all over Switzerland, to meet and shake hands with him.   First of all, Pictet asked permission of Agassiz, who answered at once : —

MONTAGNY (VAUD), 17 août, 1859.

Mon cher ami, — J'arrive de Neuchâtel et je n'ai que cinq minutes pour répondre à votre charmante lettre, avant le départ du courrier d'aujourd'hui. Je suis enchanté de vos arrangements, et je préfère infiniment que la réunion ait lieu à Genève plutôt que partout ailleurs. J'accepte avec reconnaissance votre aimable invitation pour ma femme et pour moi, et vous pouvez compter que je serai à Genthod [the splendid country house of Pictet] mardi, pour peu qu'il me reste un souffle de vie. Mille remerciements à Madame Pictet pour le bon acceuil qu'elle promet à ma femme.

Mes amitiés les plus cordials à tous nos amis communs de Genève.

Bien à vous,

LS. AGASSIZ.

The Swiss naturalists gathered promptly at Geneva and the Helvetic Society met the 24th and 25th of August in one of the halls of the Conservatory of Music. All Agassiz's old friends came in force; among them, Peter Merian of Bâle, the Nestor of Swiss geologists; Escher von der Linth of Zürich, a classmate and personal friend, and one of the first converts to the glacial theory; Bernard Studer of Berne, the explorer of the geology of the Oberland, Monte Rosa, and Grand St. Bernard; Louis de Coulon of Neuchâtel, the constant friend of Agassiz, etc., etc.; besides foreign savants of note, Tyndall and Frankland of England, Henry St. Clair Deville of Paris, and others.

The interest concentrated on the communications of Agassiz. On the first day he treated, with more than his usual brilliancy, a general question of natural history: "Toutes les grandes divisions du règne animal, telles qu'elles ont été établies par Cuvier, sont avec quelques modifications, fondées sur des bases naturelles tenant à un plan commun et non sur des bases plus ou moins artificielles," and on the second day he communicated his observations on the Acalephs, and the present formation of coral reefs on the Florida coasts in connection with an explanation of some oölitic strata of the Jura Corallian.

All the Swiss naturalists were loud in their congratulations. It was such a treat to hear again the voice of one who had made such a sensation, first at Neuchâtel in 1837, and afterwards at Porrentruy in 1838, on the subject of the glacial age and the present glaciers.

Tyndall was among the most enthusiastic, and it was a rare sight to see such savants as Jules Pictet de la Rive, Auguste de la Rive, Plantamour, A. de Candolle, A. Favre, Escher, Studer, Merian, Heer, Mousson, Dufour, Vouga, crowding round and complimenting Agassiz.

Pictet, with his usual desire to conciliate, and with the best intention, arranged, at a great party he gave Agassiz at his country seat at Genthod, a sort of accidental meeting between him and Desor; and in order to succeed in the conciliatory rôle he had undertaken, without the approbation or even knowledge of Agassiz, he tried to bring Karl Vogt to Genthod. But Vogt, who was always honest in his dealing with others, if rough and sometimes in the wrong, and who was adverse to scenes, declined to be one of the party. Agassiz was somewhat painfully impressed by the meeting, tears were shed, but, as Vogt says in his biography of Desor, no change of any sort was effected, and things remained as they were after the separation at East Boston in the spring of 1848.

Agassiz also paid a short visit to the home of his cousin, Auguste Mayor, in Neuchâtel. Although almost all the Neuchâtel families were at their summer places, they came to meet him, and he had an opportunity to see the strong feeling of friendship and admiration which all the inhabitants felt toward him. From Neuchâtel he went to Germany to visit the Braun family. His brother-in-law, Alexander Braun, had removed from Freiburg-im-Breisgau to Berlin, in 1851, and as Agassiz had not time enough at his disposal

to go so far as Berlin, Alexander arranged with his
brother Max to meet Agassiz at Aix-la-Chapelle, and
from there to go with the whole party for a two-days
visit at the house of Max Braun, who was director of
the mines at La Vielle Montagne, near Moresnet,
Belgium.   It was a meeting of congratulation on every
side, and Mrs. Agassiz and Pauline were objects of
great interest to the two brothers.   They found Agassiz
the same enthusiast, full of new schemes for the progress of natural history, and they were delighted to
learn his success in the foundation of a new museum
at Cambridge.   The moment of separation, which all
realized meant a final farewell, came only too soon,
and Agassiz left for Ostende, London, and Liverpool,
whence he sailed the 10th September for Boston.

Agassiz took advantage of his stay in Europe to purchase palæontological collections for his new museum.
He succeeded in obtaining the excellent collection of
his old teacher at the University of Heidelberg, Professor Bronn, of which he had made use when a
student there.   While in Switzerland he secured collections from the rich cretaceous localities of Ste. Croix,
Canton de Vaud, of Oeningen and Glaris, and in England he bought splendid Jurassic fossils from the
vicinity of Weymouth and Lyme Regis.   But it was
at Lièges in Belgium that he made the most important
purchase, the great collection and the no less great
library of Professor L. G. de Koninck, the author of
many monographs on the carboniferous fossils.   However, the bargain with de Koninck was not concluded
until more than a year later.

The contemplated visit to the Aar glacier and to see
the ruins of the first " Hôtel des Neuchâtelois" on the
moraine was postponed.   It was rather a disappointment to Mrs. Agassiz, who was always ready to yield
to her husband's wishes.   Although Agassiz travelled
a great deal after 1859, and Mrs. Agassiz always
accompanied him, they did not pay a second visit to
Europe, and since the death of her husband she has
not revisited Switzerland.[1]

At the end of September Agassiz returned to Cambridge, determined to spend his life in America, and at
the same time to consecrate all his energy and ability
to the creation of a great museum, according to his
own views of natural history, and, as he said to his most
intimate friends, "the best arranged and the most perfect in the world"; for in the case of Agassiz we may
apply Sydney Smith's aphorism, that "Merit and
Modesty have no other connection, except in their first
letter."

On the 14th of June, 1859, before leaving for Europe,
he had laid the corner-stone of his future great museum,
with appropriate ceremonies.   It had always been the
custom of Agassiz to start any scheme having to do
with natural history, whether publications or researches,
without thought of the necessary means to carry it out,
always confident that the future would provide the
money required.   He followed the same plan with his
museum.   He found, on entering upon his duties as
professor of zoölogy and geology in the Lawrence

[1] Lately — November, 1894 — Mrs. Agassiz left America with Mrs.
Pauline Shaw, to pass the winter in Italy.

Scientific School of Harvard University, that there were no collections in Cambridge with which to illustrate lectures upon geology and zoölogy, and that no provision had been made to obtain such collections by purchase or otherwise. Therefore, from the first day of his arrival at Cambridge, he was incessantly planning and continually adding to his private collection, sure that although not a dollar had been provided yet, and no suitable place existed, building and money would come. He himself was heavily in debt, and in addition had to provide for the daily expenses of a numerous and complicated household. Never did a man display such an amount of skilful diplomacy, — not diplomacy of the ordinary kind but natural history diplomacy, peculiar to the man, as well as peculiar to the end he wanted to attain. No one else could have succeeded, no corporation however strong and influential could have executed the plan he conceived alone and carried out alone, against all odds and constant difficulties. He never despaired of final success, although he sometimes became despondent, under pressure of illness, or of political troubles, or of dissensions among his assistants and pupils. What courage! Never was there such a valiant promoter of the progress of natural history.

The first money which came to him was the twelve thousand dollars mentioned before, raised by private subscription under the initiation and direction of the treasurer of Harvard University, Mr. Samuel Eliot, the distinguished father of the present President of the University. Then, in 1858, Mr. Francis C. Gray of

Boston left a bequest of fifty thousand dollars for the purpose of establishing and maintaining a museum of comparative zoölogy. State aid was necessary; but the question was how to get it in a commonwealth celebrated for its careful management of public money. The majority of the members of the Massachusetts Legislature are farmers, very difficult to interest in anything not directly profitable; and how to persuade them to lend the pecuniary help of the state to a purely scientific establishment, to the exclusion of similar institutions for educational purposes scattered through the state, was a problem not easy to solve; and now, more than at any other period of his eventful life, Agassiz showed of what solid metal he was made.

He first enrolled the governor of Massachusetts under his banner; then the State Committee on Education was carefully approached on general principles of public instruction, and the advantages to be derived by the farmers from a knowledge of everything relating to pests of all sorts, the best breeds of domestic animals, and kindred matters, and Agassiz obtained, by skilful manœuvring, an invitation to address the Committee on the subject. His success was now assured. What he wanted was to be brought before the Legislature; for, after privately interesting the governor and the lieutenant-governor, and the Committee on Education, he attacked the President of the Senate, the Speaker of the House of Representatives, the Secretary of the Board of Education, and the Chief Justice of the Supreme Judicial Court. In fact, the Legislature was captured, and voted that aid should be granted to the extent of

one hundred thousand dollars. Another sum, of sev-
enty-one thousand dollars, was also raised among the
citizens of Boston for the purpose of erecting a
fireproof building to receive and exhibit the collections, for which Harvard College gave a piece of
land, and Agassiz offered a plan. The plan was on
a grand scale ; the building was to be in the form of a
great rectangle open on the eastern side, the main part
364 feet in length by 64 feet in width on the western
side, with wings 205 feet in length and 64 feet in width.
It was impossible to erect such an immense fireproof
structure with the means at his disposal, and Agassiz
contented himself with building, for the present, only
two-fifths of the north wing. We shall see further on
how his plan was carried out, with great modifications,
if not in the building at least in the purpose to which
almost two-thirds of it were devoted.

In December, 1859, the condition of the north wing
was sufficiently advanced to allow the beginning of the
removal of the collections from the wooden house, near
the chemical laboratory of the Scientific School, and, in
May, 1860, the building was completed, all the collections removed, and the wooden house also changed its
place and purpose, being moved opposite the north wing
of the great museum and placed on the ground on which
the south wing was to be built later, and which is now
occupied by the Peabody Museum of Ethnography. This
small house was completely refitted and arranged as a
sort of boarding-house, called the Zoölogical Hall, for
the use only of assistants and students of the Museum.
It was a kind of third " Hôtel des Neuchâtelois," con-

ducted largely on the plan of the Pavillon Dolfuss of the glacier of the Aar. Besides the artist Burkhardt, F. W. Putnam, assistant in care of the fishes, was located there, and also successively the following students: A. Hyatt, N. S. Shaler, A. Ordway, A. E. Verrill, A. S. Bickmore, J. A. Allen, E. S. Morse, and William H. Niles.

On November 13, 1860, the inauguration of the Museum took place in the presence of the governor and his staff and escort of Lancers, and addresses were made by Governor Banks, President Felton of Harvard University, Dr. Jacob Bigelow, the chairman of the Building Committee, and Professor Agassiz. It was a happy day for the professor, and a well-deserved reward for his Herculean exertions since his arrival in the New World fourteen years previously.

Of course everything at first was in confusion. Boxes, empty or full of specimens, were piled up in all directions in the workrooms. The great library purchased from de Koninck lay on the floor in complete disorder. The labourers employed were limited to an infirm Irishman, who had great difficulty in walking, and another man absolutely inexperienced in moving specimens. Besides the cellars and attics in which were stored the collections kept in alcohol and in drawers, there were, on the first floor, four halls for the lecture room, the workrooms, and library, and on the second floor four more halls in which the collections for the public were displayed. The classification was zoölogical at first; and each assistant and pupil was assigned to a special class of animals living and fossil. Afterward the animals were divided, the fossils being in charge of special assistants, and the living animals in charge of others. To the general zoölogical classification was added what Agassiz called his "synoptic room," a sort of epitome of the whole creation, and later fauna of special geographic divisions. He even asked me to have, in a "palæontologic room," a résumé of the succession of fauna during the geological periods. After submitting a plan, which I have published since in my volume, "La Science en France," Paris, 1868, he was so enthusiastic about my plans that he wanted at oncè to put them in operation; but there was no room large enough, and to build a special hall for the purpose was too costly, and impossible at that moment.

His energy was constantly directed to innovations. Before the arrangement was made for a new placing and classification of the specimens on exhibition, he changed his plan; and this happened with him every three or four months. It was impossible to keep pace with his tremendous activity and his constant changes.

Agassiz himself was so busy that after passing from place to place, where his assistants and pupils were working, giving advice and directions, or announcing the arrival of new specimens, no time was left to him to do anything else; for every afternoon he was obliged to spend in Boston in the committee rooms of the General Court to push his claims for an annual appropriation for his Museum. Nevertheless, an immense amount of work was done every day. Each one worked with a will, for the impulse given by Agassiz was sufficient to keep every one busy.

With the opening of the Museum came a new series
of pupils. Of the old ones only four remained, —
Alexander Agassiz, Theodore Lyman, H. J. Clark, and
F. W. Putnam. Alexander Agassiz had charge of all
the specimens kept in alcohol, of the exchanges, and
the business management of the Museum, — by no
means a sinecure. Henry J. Clark had been appointed
adjunct professor of zoölogy, and his time was mainly
occupied by his lectures and his microscopic embryological studies. F. W. Putnam was in charge of the collection of fishes and vertebrates, and Theodore Lyman
worked at the Ophiuridæ.

The new set of pupils were remarkable for their
almost complete ignorance, not only of what a museum
of natural history is, but also of natural history itself.
To be sure, they all had a great desire to learn, and to
become naturalists, but they had not enough experience
to make them efficient as assistants in the work to which
they were specially detailed. On the whole, the personnel of the institution was rather crude; and the collections which came from every corner of the world became
much confused, some being determined and well labelled,
others having no labels at all or very inadequate indications. No catalogue of any sort existed, and over all
there was a mind oscillating and hesitating as regards
classification; for almost every three months, during
the first four years of the existence of the Museum,
some new idea was put forward by Agassiz, which
altered and more or less changed what he had
already proclaimed as a definite and immutable classification.

From the beginning, Agassiz adopted a very questionable method of giving his directions and instructions for the arrangement of the Museum, by calling together all his assistants and pupils into the lecture room, and there making known his plans for organization, rearrangement, classification, method, etc., explaining everything at great length and with great force.   Naturally every one applauded; and Agassiz came away from each of those numerous meetings convinced that his new directions and reforms were well understood, and that they would be put directly into operation and soon executed.   Instead of giving orders, Agassiz, with great *naïveté*, tried to convince.   The result may be easily imagined.   Every assistant and pupil after the delivery of the address, — for it was nothing more nor less than one of the usual speeches of the director, — returned to his place in the laboratories, and continued to study the work he was engaged upon, without once thinking of the advice and fresh directions just heard.   They felt a sort of inertia, which finally irritated Agassiz, who did not understand why his clear directions and instructions were not better followed.   At times he was inclined to think that it was due to the too independent American character; at other times, to ill-nature and even to conspiracies among his young pupils and assistants.   The teaching of Agassiz was unique, but the organization of the Museum was rather deficient; for the pupils, outside of their studies, had no time to arrange collections, and besides, they did not know how to proceed.   On the whole, there was considerable confusion, but no anarchy, thanks to the help and generous services constantly rendered by the friends of Agassiz, who, solely to please him, gave a great part of their time, without compensation, and often even expended liberally their own money to promote the progress of the infant establishment.

I may mention more especially the constant and generous help of Mr. James M. Barnard, an old pupil of Agassiz, who devoted a great deal of his time to the Museum, increasing its collections, working at some of them, subscribing largely to the funds of the Museum, and finally taking in hand the numerous unsold volumes and memoirs of Agassiz's works published in Switzerland, and disposing of them to the best pecuniary interest of their author. From 1855 to 1865, a period when he was always hard pressed, Mr. Barnard rendered Agassiz the greatest service in his money affairs, thus saving him from much trouble and nervous wear. To him also belongs the credit of being the prime mover and the treasurer of the Agassiz Memorial Teachers' and Pupils' Fund raised in 1874, after the death of Agassiz.

Agassiz was so earnest that it was a pleasure to help him and to increase the collections of his Museum. He was almost irresistible when he begged some favour or some beautiful and rare specimen. In the early summer of 1860, on my return from an absence of six years in Europe, I found that my friend Agassiz had much altered in his appearance and in his capacity for original study. As a teacher, he was as brilliant as ever; and as a collector of specimens, he was even more zealous than he had been; for having now a museum of his

own, he was ambitious to fill it to its utmost capacity
with all the best and rarest specimens he could obtain.
His dangerous sickness at Charleston in the winter of
1853, combined with the great mental exertions involved
in the publication of the first two volumes of his "Contributions to the Natural History of America," had told
on his strong constitution.   After mutual congratulations on meeting again,—for, although very different in
character and in our scientific researches, we agreed on
many points, and were much devoted to one another,—
Agassiz imparted to me, in his naïve way, his desire to
have me help him in his difficult undertaking.   I saw
at once the great disadvantage of creating an establishment on such a large scale, in such an out-of-the-way
place as Cambridge; but Agassiz was so sanguine and
so optimistic that it would have been cruel to raise
objections and to try to open his eyes.   It was truly
magnificent to see him every year, fighting against difficulties, especially money difficulties, with which no one
but he would have dared to contend, and always
expending double, sometimes treble, the sum he had on
hand.   Indeed, to be in debt was his normal condition.
He used to say, "Every year has its work to provide
pecuniary means for the Museum," and he acted as if
he thought he should live forever.

During the winter of 1861, Agassiz, in order to influence the Legislature and government of Massachusetts
to make a grant of money, arranged to have the governor
and the General Court to visit the Museum.   They
came in a body, and were shown all over the building,
and Mr. Barnard and I were invited to help him receive

them.   At his special request, I came to the Museum
almost daily during 1861–62, to give advice to several of
the pupils,[1] taking them on geological excursions about
Boston and Gay Head.   Finally, after helping Agassiz
gratuitously for two years, some family friends and relations in Boston subscribed a thousand dollars for two
years, to defray my travelling expenses, on the condition that I should give to the Agassiz Museum the collection of fossils made during my explorations.   In this
way I was more or less connected with the Museum
during the first four years of its existence, by far its
most difficult period; for not only was the building
small and crowded, but the space allowed for each
specialty was inadequate, the halls were cold and most
uncomfortable during the long winters, and it was not
easy even to reach the building, on account of the lack
of proper sidewalks and roads, through the surrounding
marsh.

The lack of money was a difficulty which fettered the
development of the Museum from the start.   Agassiz
had been too precipitate in his purchase of collections
in Europe, relying upon the grant of a hundred thousand dollars voted by the Legislature, the payment of
which was delayed several years; and when it was
received the trustees of the Museum refused to allow
the use of the capital, but only the income, — a great
disappointment to Agassiz, who was hard pressed for
money to pay for his foreign purchases.

[1] Agassiz recommended me more specially Messrs. Alpheus Hyatt and
N. S. Shaler, and afterward Messrs. C. Frederick Hartt and Orestes St.
John, with the request that I should help them in geology and the use of
palæontology in practical stratigraphy.

I shall give one example only to show the sort of difficulty experienced. An old scientific friend, L. Guillaume de Koninck, of Liège, on my suggestion, sold to Agassiz his splendid collection of fossils for five thousand dollars, and his no less valuable library for another five thousand dollars. Collection and library came safely over, and were unpacked during the winter of 1860–1861. But there was no money to pay for them. Agassiz, with his usual confidence, was not in the least embarrassed. " I have the collection and library, at all events," said he, "and de Koninck will have to wait." But the great Civil War came, the price of exchanges rose rapidly and so high that to send the ten thousand dollars to de Koninck would have cost fifteen thousand dollars or even eighteen thousand dollars. De Koninck was in distress, and wrote me under the date of the 19th of December, 1862 : —

J'ai toute confiance en M. Agassiz, et je crois suffisamment le lui avoir montré en lui expédiant le tout, malgré les circonstances défavorables dans les quelles se trouvent les affaires en Amérique. Comme vous avez été en quelque sorte l'intermédiaire dans cette affaire, puisque c'est d'après vos indications que je me suis adressé à votre savant ami, et que par une heureuse coincidence vous vous trouvez actuellement avec lui, vous m'obligeriez infiniment, si vous aviez la bonté de me dire vers quelle époque vous croyez que je pourrai être payé et en combien de versements? Je me suis permis de faire la même demande à M. Agassiz, mais jusqu'ici je n'ai reçu aucune réponse. Je l'ai supplié même de tacher de m'envoyer une dizaine de mille francs avant la nouvelle année, ou au moins immédiatement après, parce que j'en ai un pressant besoin pour satisfaire à des engagements souscrits. Si vous pouviez contribuer en quelque chose à me faire obtenir cette somme, qui au reste m'est due d'après nos conditions, vous me rendriez une immense service.

Another more pressing letter came soon after, and Agassiz made a settlement, agreeing to pay a yearly interest on the ten thousand dollars until he should be able to pay the whole sum. Of course, all was finally paid, but de Koninck had to wait several years.

Mr. Samuel H. Scudder has given a very good description of Agassiz's method with his students in an article entitled, "In the Laboratory with Agassiz," [1] by a former pupil, a very clever and charming reminiscence.

Mr. Scudder says : " He asked me . . . whether I wished to study any special branch. . . .   I replied that while I wished to be well grounded in all departments of zoölogy, I purposed to devote myself specially to insects.

" ' When do you wish to begin ? ' he asked.

" ' Now,' I replied.

" This seemed to please him, and, with an energetic ' Very well,' he reached from a shelf a huge jar of specimens in yellow alcohol.

" ' Take this fish,' said he, ' and look at it ; we call it a Hæmulon. By and by I will ask you what you have seen.'

" With that he left me, but in a moment returned with explicit instructions as to the care of the object intrusted to me.

" ' No man is fit to be a naturalist,' said he, ' who does not know how to take care of specimens.'

" . . . Entomology was a cleaner science than ichthyology ; but the example of the professor, who had unhesitatingly plunged to the bottom of the jar to produce the fish, was infectious ; and though

[1] " Every Saturday," Vol. XVI., pp. 369, 370, April 4, 1874, and " American Poems," several editions, published by Houghton, Mifflin & Co., Cambridge.   Also issued separately as a leaflet for the Agassiz fund, by Mr. Barnard.

this alcohol had ' a very ancient and fish-like smell,' I really dared not show any aversion within these sacred precincts, and treated the alcohol as though it were pure water.   Still I was conscious of a passing feeling of disappointment, for gazing at a fish did not commend itself to an ardent entomologist.

" . . . In ten minutes I had seen all that could be seen in that fish. . . .   Half an hour passed, an hour, another hour; the fish began to look loathsome.   I turned it over and around; looked it in the face, — ghastly! from behind, beneath, above, sideways, at a three-quarters view, — just as ghastly.   I was in despair.   At an early hour I concluded that lunch was necessary; so, with infinite relief, the fish was carefully replaced in the jar, and for an hour I was free.

" On my return, I learned that Professor Agassiz had been at the Museum, but had gone, and would not return for several hours. . . .   Slowly I drew forth that hideous fish, and, with a feeling of desperation, again looked at it.   I might not use a magnifying glass; instruments of all kinds were interdicted.   My two hands, my two eyes, and the fish; it seemed a most limited field. . . . At last a happy thought struck me — I would draw the fish; and now, with surprise, I began to discover new features in the creature. Just then the professor returned.

" ' That is right,' said he; 'a pencil is one of the best eyes.   I am glad to notice, too, that you keep your specimen wet and your bottle corked.'

" With these encouraging words, he added : —

" ' Well, what is it like ? '

" He listened attentively to my brief rehearsal of the structure of parts whose names were still unknown to me. . . .   When I had finished, he waited as if expecting more, and then, with an air of disappointment, ' You have not looked very carefully; why,' he continued most earnestly, ' you haven't even seen one of the most conspicuous features of the animal, which is as plainly before your eyes as the fish itself.   Look again! look again!' and he left me to my misery.

" I was piqued; I was mortified.   Still more of that wretched fish!   But now I set myself to my task with a will, and discovered one new thing after another, until I saw how just the professor's criticism had been. The afternoon passed quickly, and when, toward its close, the professor inquired, —

"'Do you see it yet?'

"'No,' I replied, 'I am certain I do not; but I see how little I saw before.'

"'That is next best,' said he earnestly; 'but I won't hear you now; put away your fish and go home; perhaps you will be ready with a better answer in the morning. I will examine you before you look at the fish.'

"This was disconcerting. Not only must I think of my fish all night, studying, without the object before me, what this unknown but most visible feature might be, but also, without reviewing my new discoveries, I must give an exact account of them the next day. I had a bad memory, so I walked home by Charles River in a disturbed state with my two perplexities.

"The cordial greeting from the professor the next morning was reassuring. Here was a man who seemed to be quite as anxious as I that I should see for myself what he saw.

"'Do you perhaps mean,' I asked, 'that the fish has symmetrical sides with paired organs?'

"His thoroughly pleased 'Of course, of course!' repaid the wakeful hours of the previous night. After he had discoursed most happily and enthusiastically — as he always did — upon the importance of this point, I ventured to ask what I should do next.

"'Oh, look at your fish!' he said, and left me again to my own devices. In a little more than an hour he returned and heard my new catalogue.

"'That is good, that is good,' he repeated; 'but that is not all; go on.' And so for three long days he placed that fish before my eyes, forbidding me to look at anything else or to use any artificial aid. 'Look! look! look!' was his repeated injunction.

"This was the best entomological lesson I ever had, — a lesson whose influence has extended to the details of every subsequent study; a legacy that professor has left to me, as he left it to many others, of inestimable value, which we could not buy, with which we cannot part.

" . . . The fourth day a second fish of the same group was placed beside the first, and I was bidden to point out the resemblances and differences between the two ; another and another followed, until the entire family lay before me, and a whole legion of jars covered the table and surrounding shelves. The odour had become a pleasant perfume ; and even now the sight of an old six-inch, worm-eaten cork brings fragrant memories.

" . . . Agassiz's training, in the method of observing facts and their orderly arrangement, was ever accompanied by the urgent exhortation not to be contented with them.

" ' Facts are stupid things,' he would say, 'until brought into connection with some general law.'

" At the end of eight months, it was almost with reluctance that I left these friends and turned to insects ; but what I had gained by this outside experience has been of greater value than years of later investigation in my favourite group."

Agassiz delivered lectures twice or three times a week before his students. It was the custom at the Museum for every one, assistants, as well as pupils and friends, to attend. His wife was always present, taking notes most faithfully of all that he said. Never was there such a devoted secretary. The numerous public lectures in America, delivered during fifteen years before large audiences, exerted an influence easy to perceive. His explanations were clear, right to the point, never too scientific, and, as far as possible, in untechnical words. But he wanted applause, and even courted it, with great skill. Like an actor on the stage, he would pause at the end of a sentence, in order to allow time for applause. His lectures then and after took more the form of addresses than ordinary expositions of scientific questions. He always interested his audi-

ence to a high degree, but his best teaching was not
done in this way.   It was too theatrical.   Agassiz
was at his best when he was looking at specimens,
detailing the differences, dwelling strongly on characters which would have escaped other eyes than his own.
His unusually keen sight and his rare memory were his
great attractions, and he displayed them to the best
advantage, on the spur of the moment, when there
were only one or two hearers.   Then Agassiz was
natural, without thought of effect; and he was a wonderful and rare naturalist and incomparable teacher.
Like many men of genius, he would seek for admiration and applause, even in dealing with questions, or
more correctly with branches, of science, upon which
he was not well informed.   Trusting to his great ability and immense experience in lecturing, he would go
deeply into a subject, not only foreign to his usual
researches, but which he had really no inclination to
investigate fairly; like the great French painter Ingres,
who preferred the approval given to his rather indifferent performance on the violin to that accorded his great
and splendid pictures.

Agassiz was not a practical geologist; and when in
the field, he showed an almost complete absence of
the intuition requisite, indeed absolutely necessary, to
master the stratigraphy, the classification, and the orog-
raphy of any portion of the earth's surface — excepting
in regard to glacial questions, on which he was a great
master.   In his lectures he liked to go into historical
geology, in which branch his knowledge was absolutely
defective.   So long as he kept to generalities, he did

very well; but as soon as he entered into the details of strata, he was weak, wholly lacking in exactness, and uninformed as to geographical geology and the sequence of the numerous groups into which the strata have been divided in each country.   If I happened to be present at such lectures, — which was very sel- dom the case, — Agassiz would look at me in a ques- tioning and imploring way, which said as plainly as words, "Please do not contradict me."   Of course I never did.

As a résumé of his unequalled capacities and talents as a teacher, we may repeat what was said of the great French geometer Monge: "Il combinait pour la clareté de ses démonstrations, les regards, les paroles et les gestes.   Ses auditeurs craignoient de faire le moindre mouvement dont le bruit put troubler le charme de cette étonnante éloquence."

Unhappily difficulties of another nature than money stringency occurred in the Museum; namely, with the personnel of the establishment.   Agassiz never knew how to manage his assistants, and repeated the same faults, with some little variation of details, in his Cam- bridge Museum, which he had committed previously at Neuchâtel.   At first all went smoothly.   As a sort of re- ward, he sent most of his assistants to the Smithsonian Institution at Washington, to study its management and arrange exchanges of specimens.   The first difficulty was with Professor Clark, who left the Museum in June, 1863, as we have seen.   Then came a sort of revolt among most of his other assistants, which developed slowly during the years 1863 and 1864, and broke out at the begin-

ning of 1865. A regular secession occurred, and five
assistants left. One, Mr. Verrill, went to the Smithsonian and afterwards to Yale College; while the others,
Messrs. Putnam, Morse, Packard, and Hyatt, retired to
Salem, the county seat of Essex County, as a sort of
Mount Aventine, where George Peabody, an American
banker in London, had given a certain sum of money to
the Essex Institute and Peabody Academy of Science.
And around these Agassiz's four pupils collected, using
them as a base or citadel, from which they expected to
conquer the natural history of North America; and for
this purpose they started a monthly magazine, "The
American Naturalist" and an agency for selling scientific books and papers and for exchanging specimens.
After a few years of hard struggle the enterprise failed,
notwithstanding their residence among people whom
the novelist, W. D. Howells, satirically characterizes as
"a little above the salt of the earth." Two were glad
to return to Cambridge, one went to Brown University, Providence, while one has remained at Salem,
as curator of the Peabody Museum and of the Essex
Institute. The whole disagreement was a great mistake on both sides, but more especially on the part
of the pupils, who ought to have been more patient
with their old professor. The crisis was brought on
by new regulations for the assistants of the Museum.
Agassiz, with justice, requested that no one connected
with the Museum as a regular assistant, or even as a
student-assistant, should work for himself in the Museum
during the hours fixed for Museum work. He contended
that persons were nothing, and that the Museum was

above everything and every one. It was, perhaps, a little too much to expect that young men, who had come there to learn and fit themselves for positions as teachers and professors of natural history, should give the best of their time every day to Museum work, especially in view of the fact that the majority of them were not paid at all. All had come with expectations of situations and reputation in the near future; and besides this, almost all possessed that independence of character peculiar to Americans, and more particularly to New Englanders. Agassiz was at first too lenient, and afterward became too exacting, while the liberty and equality existing there were not conducive to good discipline. In fact, all the regularly appointed assistants or student-assistants were working for themselves as much as they could. The curator had too much in hand to see that every one was doing his duty. Besides, no definite duties were assigned to most of the assistants; and if one was given the superintendence of a department, he had no one to help him, even in moving and carrying specimens, writing labels, or cataloguing and numbering the collections.

On the other hand, the assistants and pupils should have considered that it was a great privilege, and at that time an opportunity unique in all North America, to help in building up a great museum under the leadership of a naturalist of genius, and that they were there, not only for their own instruction, but also for the good of future generations. Egotism played too large a part with them, and caused Agassiz great disappointment. He had become very irritable through overwork, and signs of failing health were only too visible. His young assistants ought to have been more considerate, and endured with more composure the weak points of their position. Both sides suffered by the separation, but the assistants suffered the most; and to this day the result of this secession is still visible on most of them, although they are all very proud of being old pupils of Agassiz.

CHAPTER XVIII.

1858–1864 (continued).

DARWIN'S "ORIGIN OF SPECIES" — CUVIER, AGASSIZ, OWEN, LAMARCK, AND DARWIN — THE OPPONENTS OF AGASSIZ IN AMERICA — ASA GRAY AND CHAUNCEY WRIGHT — PARALLEL BETWEEN CHAUNCEY WRIGHT AND KARL SCHIMPER — TWO CLASSES OF NATURALISTS — REVOLUTION AND EVOLUTION — PIETIST AND ATHEIST — LYELL'S DISSENT — NEO-LAMARCKIANS AND NEO-DARWINIANS — UNIFORMITARIANISM — SPONTANEOUS GENERATION — TRUE POSITION OF CUVIER AND AGASSIZ.

DARWIN'S "Origin of Species" appeared in London, the first of October, 1859, a few days after Agassiz's return from Europe. The book was a disappointment to almost all naturalists. It had been heralded several years before it was issued as an event which was likely to revolutionize natural science, and so it did. Our friend Sir Charles Lyell more especially was enthusiastic and mysterious about it; but six months after its publication, when I visited him in London, I found him rather more frightened than pleased. He says, " Darwin goes too far. I am not prepared to follow him on natural selection and the geological succession of organic beings; the last more especially is still too incomplete to draw such sweeping conclusions."

In England adherence to, and even sympathy with,

Darwin's essay was rather limited.   Besides Lyell and
Wallace (then at Ternate, Molucca Isles), who accepted
the new doctrine with reservations, there were only
the botanist, Sir Joseph D. Hooker, the anatomist and
essayist, Thomas H. Huxley, and two entomologists,
H. W. Bates and Sir John Lubbock, who could be called,
from the first, friends of the theory of the descent and
modification of species according to Darwin's views.
All the English palæontologists, with Sir Richard
Owen at their head, were opposed to it; and not long
after the appearance of the " small green-covered book,"
as it is called by Huxley, Owen in his library, at the
Sheen Lodge, Richmond Park, used to make fun of
Darwin's modification of species under domestication;
and pointing to his pet dog, as he tried to catch flies,
he would say, " My dog is in the act of becoming one
of the insectivora."[1]

Darwin's old teacher at the Cambridge University,
Professor Adam Sedgwick, was anything but pleased
on receiving a presentation copy of the " Origin of
Species," which he read " with more pain than pleas-

[1] Attempts have been made, not only to question the influence exerted
on Richard Owen by Cuvier, but to place Owen among the Darwinists,
and even to call him a precursor of Darwin in the much-controverted
question of " Natural Selection."   Owen, like Agassiz, was truly a disciple
of Cuvier.   Owen until the end of his life always uttered the name of
Cuvier " with grateful reverence "; and in 1883 he wrote to Georges
Frédéric Cuvier, nephew of Georges and son of Frédéric: " There are
fashions of thought as well as of dress.   A somewhat prevailing one, to
which you allude, I have occasionally referred to as the *Biologie con-
jecturale :* but the science of living things which will endure is based on
the foundation of the *faits positifs* made known, with the true methods of
their discovery, in the immortal works of Georges Cuvier " (" The Life
of Richard Owen " by Rev. R. Owen, Vol. II., p. 249; 1894).

ure. Parts of it I admired greatly, parts I laughed at
till my sides were almost sore; other parts I read with
absolute sorrow, because I think them utterly false and
grievously mischievous. Many of your wide conclusions are based upon assumptions which can neither
be proved nor disproved. . . . Darwin has deserted
utterly the inductive track and taken the broadway of
hypothesis" ("Life and Letters of Adam Sedgwick,"
Vol. II., pp. 356–357; Cambridge, 1890). Other celebrated English naturalists, like A. Murray, J. E. Gray,
Harvey, and Henslow, would have none of it.

On the continent of Europe, the only open sympathizers were two botanists, Alphonse de Candolle of
Geneva and Charles V. Naudin [1] of Paris; and as a
matter of course, the learned and always most progressive geologist, J. J. d'Omalius d'Halloy, the steadfast
friend and pupil of Lamarck, but he was rather sur-

[1] Previous to Darwin's publication, the French botanist, Naudin, had
made some remarkable researches on the hybridization of plants; but all
his efforts to penetrate the mechanism of variation of species had been
unsuccessful. Another French botanist, Joseph Decaisne, desirous to
prove for himself the question of variations, made a series of experiments,
lasting more than twenty years, on the genus *Pirus*. He chose a rare
example of cultivation of the pear tree, at the Garden of the Chartreux,
Palace of Luxembourg, Paris, which had continued without interruption
during several centuries. The number of typical species or varieties of
pear trees cultivated there had reached, in 1871, the large number of
fourteen hundred. As a result of his experiment, Decaisne arrived at the
unexpected conclusion that all pear trees belong to an *unique type*,
although a polymorph one; and writing to a friend in 1868, he said, " Si
la nature n'a pas employé d'autre procédé [referring to the natural selection of Darwin] pour façonner le monde actuel, il ne doit pas être difficile
de la prendre sur le fait," to which he added, " Je voudrois voir cela de
mes yeux." Here lies the real difficulty, — the impossibility of seeing
with one's own eyes the arrival of a new species.

prised to find that Darwin had almost entirely ignored
Làmarck, as if he had never existed. For d'Omalius
the "Origin of Species" was the "Philosophie Zoologique" of Lamarck in a new dress.

In America it was different. Under the leadership
of the botanist, Asa Gray, and the geologist, William B.
Rogers, almost all the naturalists became at once strong
Darwinians. Agassiz tried in vain to stop the sweeping
wave, but he was overwhelmed by the flood of publications and reviews. What was rather annoying to him
was that the most enthusiastic propagators and apostles
of the new gospel were not naturalists at all, with the
exception of the systematic botanist, Asa Gray. Not
one of them was a zoölogist, in any sense of the word.
Agassiz was too much a naturalist to accept a number
of mere suggestions until they were scientifically proved
by exact observations. In his eyes Darwin was an
advocate of a foregone conclusion, who argued, not for
the purpose of finding in what direction the evidence
of any particular fact would lead, but for the purpose
of finding something in the fact favourable to his
preconceived opinion. Agassiz himself had had the
honour to overthrow too many errors and false generalizations, not to be open to all new facts and investigations. But where were the facts? Darwin admitted
the difficulties in his theory, which he tries to explain
away, not by well-grounded facts and careful observations, but by various suppositions and many *ifs;* and
these through frequent repetition seem to become established truths in his mind, and are used as arguments.
On the contrary, Agassiz had gathered together, dur-

ing his many years of close investigations, a mass of facts which were not favourable to Darwin's somewhat hasty conclusions, and more especially to those of his followers, who at once exaggerated many of his views and conclusions.

This controversy brought to Agassiz's mind the great discussion between Cuvier and Geoffroy Saint-Hilaire, at the meetings of the Academy of Sciences of Paris, in 1830.[1] Cuvier, who was the greatest debater natural history has ever had, with his prodigious memory, had every fact at his tongue's end, and was always able to accumulate such a mass of proofs against an adversary that it was useless to oppose him. Although in this case Geoffroy Saint-Hilaire was right, as it has been amply proved since, he was fairly defeated in each day's encounter with Cuvier, and withdrew, so overpowered by his great opponent that he tottered like a drunken man, not knowing where he was, nor where he was going, so one of the witnesses of those discussions has told me.

Agassiz, although a rare teacher and a remarkable

---

[1] It began on the 22d of February, 1830, and was occasioned by a report made by Geoffroy on a paper, on the Organization of the Cephalopods, written by two young and obscure naturalists. In his report, Geoffroy advanced his new views on the unity of organic composition and unity of types — the result of more than thirty years of constant research. The discussions, which lasted, with one interruption only, during the whole year, are well summarized in the book published in May, 1830, by Geoffroy, under the title " Principes de Philosophie Zoologique," which may be considered as a basis for the classification of all the facts of comparative anatomy; and also in the chapter, " Discussion Académique de 1830," in the " Vie, travaux, et doctrine scientifique d'Étienne Geoffroy Saint-Hilaire," par son fils Isidore, pp. 366–385. Paris, 1847.

lecturer, was a poor debater. While Cuvier always
kept cool and carefully selected every word he used,
Agassiz, on the contrary, quickly lost patience, became
excited, and showed signs of vexation. It should be
said, however, that this was not the case with Agassiz
until after he was fifty years old. I remember to have
seen him very cool and under perfect self-control in
several of the discussions on the glacial theory before
von Buch, Élie de Beaumont, and other adversaries of
his glacier theory, who on their part to a greater or less
degree lost their tempers. At the meetings of the
American Academy of Sciences and of the Natural
History Society of Boston in 1860, objections against
the acceptation of Darwin's theory led to several debates
between him and Asa Gray and William B. Rogers, in
which he was defeated, although he was right in all
the facts he advanced to sustain his views and opinions.
Some of his antagonists were excellent debaters and
skilled in interrupting; and they annoyed him constantly
by shaking their heads, or even saying a few words aloud,
which disconcerted him and produced a painful impression. His opponents have reproached him for taking all
that they said as directed against him personally when
they were only making objections to his arguments and
the views he expressed and defended. Probably the
reproach is just, to a certain extent; but that Agassiz
should take it as a personal opposition is easy to understand, and was at least partly authorized by certain
facts. His great success and popularity in America
had arrayed against him almost all the American naturalists, with very few exceptions. Even his position at

Harvard University was considered by some as unjustly bestowed on him, a criticism which was particularly applicable in regard to his geological work ; and it was current among a certain public, happily a rather limited number, that the merits of Agassiz had been "much overrated in America." It would have been politic on his part if he had offered the chair of geology to William B. Rogers, then a resident of Boston. But Agassiz did not like to have any one so near, who might overshadow him.

But, however it happened, Darwin's "Origin of Species" became a thorn in his side. His pupils in a body turned against him, for they were delighted to believe that they knew more than he of the philosophy of natural history, the descent of man, the creative power of horticulturists, and of pigeon breeders, and the mutability of species and genera. To the disgust of Agassiz, they turned from their master to applaud all the articles on evolution and origin of species, published in American periodicals by Asa Gray,[1] Chauncey Wright, and John Fiske, the last two not even naturalists.

[1] Almost a year before the publication of Darwin's "Origin of Species," Gray, in January, 1859, read before the American Academy of Arts and Science, a paper in which, as he said in a letter to Torrey, he "knocked out the underpinning from Agassiz's theories about species and their origin, showing by the very facts that threw de Candolle, the high probability of single and local creation of species, turning some of Agassiz's own guns against him" ("Life of Asa Gray," Vol. II., p. 450). It is plain that Darwin's book came just in time for Gray, who seized upon it at once, and used Darwin's weapons against Agassiz with a quickness, which was not free from some passionate opposition. To be sure, Lamarck's "Philosophie Zoologique" was within Gray's reach, but it is

A few words on Chauncey Wright, and his singular
similarity to another adversary of Agassiz, Karl Schimper, will not be out of place.   Chauncey Wright was a
mathematician of talent, who turned his mathematical
skill to a study of the phyllotaxis of plants, just as
Schimper had done forty years before.   Agassiz treated
Wright in the most friendly way, even appointing him
a lecturer at his school for girls, just as he had treated
Karl Schimper.   Wright was an earnest seeker for
truth, but he was above all a great dreamer, and some
of his writings are rather obscure.   He was suffering
from the same weakness which afflicted Schimper, and
presents a rare parallel to him.   As his biographer and
friend, Mr. Charles Eliot Norton, says, "He was never a
persistent and systematic student, but he was essentially
a persistent and systematic thinker" ("Philosophical
Discussions," p. xvi., 1877; New York).

Asa Gray "had no proper training in biological
science," Huxley says, and this was certainly true of
Chauncey Wright, and especially of John Fiske.   All
three were ignorant of zoölogy, and it was almost comical to have Wright say, "Darwin's 'Origin of Species'
renders Agassiz's essay on classification a useless and
mistaken speculation; creation is a word pretending
knowledge and feigning reverence."

In an address of Professor Asa Gray[1] on Professor
Jeffries Wyman, we read, "I may venture to take the

doubtful if he had ever read it, not being very proficient in the French
language.   At all events, Gray's attention was not called to Lamarck's
work until after the publication of Darwin's "Origin of Species."

[1] "Proceedings Boston Soc. Nat. History," Vol. XVII., p. 123.
1874.

1858–64.]   *JEFFRIES WYMAN.*   111

liberty to repeat the substance of a conversation which
I had with him [Jeffries Wyman] some time after the
death of the lamented Agassiz, and not long before his
own. I repeat the substance only, not the words."

"Agassiz repeated to me," he said, "a remark made
to him by Humboldt, to the effect that Cuvier made a
great mistake, and missed a great opportunity, when he
took the sides he did in the famous controversy with
Geoffroy Saint-Hilaire. He should have accepted the
doctrine of morphology, and brought his vast knowledge
of comparative anatomy and his unequalled powers to
their illustration. Had he done so, instead of gaining
by his superior knowledge some temporary and doubtful victories in a lost cause, his pre-eminence for all
our time would have been assured and complete. I
thought," continued Wyman, "that there was a parallel
case before me — that if Agassiz had brought his vast
stores of knowledge in zoölogy, embryology, and palæontology, his genius for morphology, and all his quickness of apprehension and fertility in illustrations, to the
elucidation and support of the doctrine of the progressive development of species, science in our day would
have gained much, some grave misunderstandings been
earlier rectified, and the permanent fame of Agassiz
been placed on a broader and higher basis even than it
is now."

These opinions of Wyman, quoted and indorsed by
Gray, indicate an inclination in both to say, "what a pity
that Cuvier and Agassiz did not at once accept Lamarck's, Geoffroy Saint-Hilaire's, and Darwin's theories
of descent; if they had done so, they would have been

greater men." On the contrary, both would have in-
jured their record as exact observers and true savants.
Their convictions were based on researches in laboratories, which had lasted all their long lives, and it would
not have been to their honour to give up all the facts
they had patiently accumulated, in order to adopt views
contrary to what they had seen and observed. Both
Cuvier and Agassiz were very honest, and had too high
an idea of their priesthood in natural history, not to
protest against the acceptation of theories not fully sustained by facts patiently accumulated. Darwin, in a
letter to Sedgwick, says, that his volume on the "Origin
of Species" was the result of more than twenty years'
study, during which he "worked like a slave on the
subject." But Cuvier also worked like a slave during
forty years, and so did Agassiz.

Naturalists may be divided into two categories : those
who are philosophical naturalists, and those who are,
above all, guided by well-observed facts. Philosophers
are all dreamers and isolate themselves as much as they
can, not only from society, but even from companionship with their fellow-workers. Lamarck, although
suffering from weakness of the eyes, and finally becoming blind, led an isolated life ; he was not sociable even
with his colleagues in Paris ; and Pyramus de Candolle, who became his collaborator in the third and
fourth edition of his "Flore Française," did not affiliate
with him at all, while, on the contrary, he was intimate
with, and a great admirer of Cuvier. Geoffroy Saint-
Hilaire also became blind. Naudin was completely deaf.
Darwin's constant suffering and complete isolation at

his country house, " Down, Bromley, Kent," are well known. We read in his life, Vol. I., p. 243 : " Those two conditions — permanent ill health and a passionate love of scientific work for its own sake — determined thus early in his career the character of his whole future life." An avowal of Huxley to the son of Darwin may also be added : " Like the rest of us [which means Joseph Hooker, Asa Gray, Charles Lyell, and himself], he [Charles Darwin] had no proper training in biological science." Bates passed eleven years in the Amazon valley, cut off from all scientific society and absolutely isolated ; and when I saw him in London, in 1870, he was living like a hermit. Wallace is another example of a traveller who lived year after year in the naturalists' paradise of both the new and old worlds : four years on the Amazons and eight years in the Indio-Malayan archipelago. During their long isolation, naturalist-philosophers are apt to theorize ; more especially if, like Bates and Wallace, they start with the avowed purpose of finding the origin of species. The weak points seem to them only imperfections in the records, which will be filled up by and by, and each believes that he has found the laws of variation and of evolution in the organic world.

The second class of naturalists, who may be called classifiers and pioneer-naturalists, do not isolate themselves, and are anything but hermits. They work in laboratories as well as in the field, always well equipped and drilled to observe every organism ; and they are disinclined to theorize, until all the facts lead them toward an inevitable conclusion. They are constant attendants at scientific meetings of academies and natural history societies, discussing there every new fact brought up by any observer, whatever may be his scientific position and record. They try to classify the immense stores of collections around them, critically examining every specimen, and observing without prejudice everything that comes under their keen eyes.

The majority of them never indulge in dreams; or, if they do, they do not allow their imagination to go beyond the limit of speculation which can be easily seen and readily controlled by the immediate researches of their fellow-workers, as well as their own. To this category belong Cuvier, Agassiz, Owen, d'Orbigny, Deshayes, Ed. Forbes, Thomas Davidson, Pictet de la Rive, Herman von Mayer, Barrande, Lartet, Cotteau, and almost all palæontologists. Cuvier was unique for his constant adherence to facts. He never dreamed in his life. Agassiz, next to him, was influenced only by facts, though he dreamed now and then; but his great practical experience in both hemispheres soon put an end to all wild conclusions or hasty speculations which may have come now and then to his impetuous spirit, and brought him back face to face with the simple facts. Which of the two classes of men do the best work in natural history? is a question easy to answer. All true and solid progress is due to the second class and without them natural history could hardly be comprehended. Philosophical naturalists would find their task a very barren one if there were no classifiers, no embryologists, no palæontologists. It is very well to theorize and discuss teleology, agnosticism, spiritism, morphology, mimicry, natural selection, evolution, transformism, etc., but before everything else we must know the history of every animal, of every plant, and accumulate all that constitutes the treasuries of every branch of natural history.   Notwithstanding the saying of Bates that " Darwin and Hooker have elevated natural history into the rank of an *inductive science,* instead of being only *the observation and cataloguing of facts*" (letter to Dr. J. S. Hooker, March, 1861), it is difficult to decide how Bates would classify the comparative anatomy of Cuvier, the stratigraphical system of organized fossils of William Smith, or the Ice period of Agassiz.

   There are two words which have an almost supernatural influence on the naturalists of the nineteenth century, — a circumstance which is easily explained when we consider how far humanity is led by words, and that fashion exists in everything human. During the first part of this century, the word "revolution" was extensively used in natural history.   It was natural for persons who had witnessed the great French Revolution of 1789 to liken all that was extraordinary and difficult to explain in natural history to great revolutions.   So Cuvier, with his "Revolutions de la surface du globe," started a whole literature.   Everything was revolution and catastrophe.   Humboldt spoke of revolutions in both hemispheres; Élie de Beaumont spoke of revolutions and elevations of mountain systems;. de Boucheporn revolutionized the theory of the earth by his explanation that the changes in its axis were due to

the shock of comets, etc., etc. Later, since October,
1860, when Herbert Spencer, in his "First Principles of
a New System of Philosophy," gave the following definition : " Evolution is the integration of matter and con-
comitant dissipation of motion," everything in natural
history, and in social, political, economical, historical,
and philological sciences, is evolution. It is not that the
word "evolution" was not used before in natural history, although, curiously enough, Darwin did not use it
once in his "Origin of Species"; for since the second
half of the eighteenth century we find it used by
Bonnet, and afterward by Laurillard, the assistant and
right hand of George Cuvier, who recalled the observations of Bonnet on evolution. If Lamarck had used it
in his "Philosophie Zoologique," his theory might have
had another destiny during his lifetime. Darwin had
the acuteness to see what a capital handle it would
make for his theory, and as soon as he saw it in Spencer's work, he transferred the word into all his other
works, speaking constantly of the "principles of evolution." His sympathizers took to evolution, and now
evolution is everywhere. It has dethroned revolution
completely. The word "evolution" *a fait fortune* according to a French proverb.

It is the only tie — certainly a very slender and
elastic one — between all those who call themselves
Darwinians ;[1] although the word does not occur at all
in the "Origin of Species." In this connection a few
quotations from "Asa Gray's Life and Letters" are

---

[1] There is only one exception, Alphonse de Candolle, who used the
word "transformism" as preferable to "evolution," because, as he says,

interesting. He says, "Lyell considers the case [transmutation question] as not yet ripe for a decision." "Lyell does not come out as an advocate of natural selection, transmutation" (both Darwin and Hooker complain of it). "Lyell has presented the case of transmutation so as to commend it as much as possible to us orthodox people. [Huxley would have put it in a way to frighten us (orthodox) off.]" Here is the great obstacle between Gray and Huxley, and even Darwin. Gray says, "As to the Exeter meeting of the British Association, I am, on the whole, glad enough to keep away, especially from Darwinian discussions, in which I desire not to be at all *mixed up* with the prevailing and peculiarly English materialistic, positivistic line of thought, with which I have no sympathy, while in natural history I am a sort of Darwinian" ("Letters of Asa Gray," Vol. II., p. 592). And in two other places Gray says: "In Darwin's contributions to teleology, there is a vein of *petite malice*, from my knowing well that he rejects the idea of design." "You (Darwin) see what uphill work I have in making a theist of you of *good and reputable standing*." Darwin is in favour of chances, while Gray feels a "cold chill" when Darwin brings him to the point of co-adaptations in orchids. Certainly the association of an orthodox Presbyterian, or strict Puritan of the old school, like Gray, with the agnostic and materialistic Darwin and Huxley is curious. Circumstances made Gray a Darwinian, for

the successive changes of forms do not occur always in the direction of a greater development, being sometimes in the direction of simplification or even of the production of monstrosities ("Darwin," 2d edition, p. 35; Genève, 1882).

he always stopped short of complete deduction. For instance, when he says, "in Bates's geographical varieties,
etc., we get about as near to seeing a species made as we
are ever likely to get." Darwin and Huxley do not think
so. Gray leaned towards Owen's definition, that species
are somewhat derived genealogically, and he urges Darwin "not to insist much on natural selection, if you can
only have derivation of species"; and adds, "derivation
of species is to be the word." Finally, Gray accepts
the conclusions of Darwin as a "probable hypothesis."

As to Lyell, he says in a letter to Darwin, "I cannot
go Huxley's length in thinking that natural selection
and variation account for so much, and not so far as
you, if I take some passages of your book separately.
I think the old 'creation' is almost as much required
as ever, but, of course, it takes a new form if Lamarck's
views improved by yours are adopted" ("Life of
Charles Lyell," Vol. II., p. 363).

In another letter also to Darwin, he says, "Lamarck's
belief in the slow changes in the organic and inorganic
world in the year 1800 was surely above the standard
of his time, and he was right about progression in the
main, though you have vastly advanced that doctrine"
("Charles Lyell," Vol. II., pp. 365, 366). And farther on
Lyell entirely breaks with some of the main conclusions
of Darwin, as when he says, "I feel that progressive
development or evolution cannot be entirely explained
by natural selection. I rather hail Wallace's suggestion that there may be a Supreme Will and Power
which may not abdicate its functions of interference,
but may guide the forces and law of Nature. . . .

At the same time, I told Wallace that I thought his arguments as to the hand, the voice, the beauty and the symmetry, the naked skin, and other attributes of man, implying a preparation for his subsequent development, might easily be controverted " (" Charles Lyell," Vol. II., p. 442).

Lyell is the only Darwinian who has made any reference to "spontaneous generation." It is in a letter to Charles Darwin, dated March 15, 1863 (" Charles Lyell," Vol. II., p. 346). Curiously enough, he calls Richard Owen " a disciple of Pouchet " of Rouen. Darwin and Huxley have gone as far as it was possible for them to go, in reducing the initiative beings on earth to four or five cells, even to a single one, according to Darwin's most intimate thought. From that cell to spontaneous generation there was an easy passage, especially for materialists and agnostics. But the very exact and splendid experiments of Louis Pasteur, proving beyond discussion that "spontaneous generation" does not exist, broke the first link of their chain of reasoning for the " Origin of Species," and they all by common consent passed over it as too dangerous ground.

Gray's slight knowledge of geology, palæontology, and zoölogy led him to overestimate the value of both Lyell and Darwin, when he says, " It is interesting to see how early he [Lyell] took to the line which he followed in his whole life's work, and which has changed the face of geology and philosophical natural history. For, indeed, Lyell is as much the father of the new mode of thought which now prevails as Darwin " (Letter of Gray to A. de Candolle in " Letters of Asa Gray,"

Vol. II., p. 732). The influence of Lyell on geology, outside of a very limited circle in the British Isles, is absolutely insignificant. Geology has been built up entirely outside of his "line of work," by practical geologists and original observers, as well as by great thinkers. That Lyell was a charming character is true, and it is also true that he influenced Darwin, but he did not in the least change the face of geology.

The publication of some letters of Darwin, Lyell, and Gray, in their respective biographies, although carefully selected, apparently to do honour to their authors, shows in part the inside history of the strong divergences existing among evolutionists and uniformitarians. Both Lyell and Gray feel "cold chills" every time they come to the full conclusions drawn by Darwin, Huxley, and Wallace. They had to come face to face with agnostic and antitheistic ideas. The bridge they had to cross seemed to both of them too insecure to trust their feet on it, just as Agassiz and many others refused to cross the bridge of the "Origin of Species" of Darwin.

Some of the quotations from the letters of Darwin are certainly amusing. He calls Hooker "a barrister, a great lawyer"; Lyell is "a Lord Chancellor"; Asa Gray is "a born reviewer, a capital reasoner, a poet, a hybrid, a complex cross of lawyer, poet, naturalist, and theologist! was there ever such a monster seen before?" And he wrote to Asa Gray, "You have made a mistake in being a botanist; you ought to have been a lawyer" ("Darwin's Life," Vol. II., p. 120). This is true. Lyell, Hooker, and Gray were certainly lawyers,

and all three constantly applied lawyers' methods to
natural history.   No one of them possessed the naturalist spirit and turn of mind.

Of Joseph Prestwich, Darwin says, " I fear he is too
much of a catastrophist."   For him " Huxley is a
regular reviewer."   Never a man was more inclined to
paradoxes than was Darwin.   For instance, he says,
" A compiler is a great man, ·and an original man, a
commonplace man";   "only fools can generalize and
speculate " (in a letter to Hooker).   His treatment of
Lamarck's views and observations do not speak in his
favour, for in one place he says that Lamarck's work
appears to him " extremely poor "; and again, " I got
not a fact or idea from it."   Lamarck's book is " ab-
surd, an absolutely useless book," and all that Lamarck
did is " rubbish."   However, he admits that Lamarck is
the only exception among all those who have described
species, for he did not believe in the immutability and
permanence of species.

Darwin applied the word "rubbish" rather at random,
and not always to the point.   For instance, he characterized by that title the great discovery of the antiquity
of man by Boucher de Perthes at Amiens, and a few
months later he scolds Lyell for not rendering sufficient justice to Boucher de Perthes in his book " The
Antiquity of Man," saying that " Boucher de Perthes
has done for man something like what Agassiz did for
glaciers."

On the whole, Darwin was a great sceptic, and a
lover of paradoxes, full of preconceived ideas, although
he constantly protested that he had never had any, ex-

cept one, and that one was very unfortunate, for it was
in regard to the formation of atolls by corals, which he
preconceived when on the coast of South America,
before he had seen a single coral island of Australia.
Since 1874 Darwin's theory has been so often damaged
by the numerous facts brought forward by Semper,
John Murray, Agassiz, Guppy, and Bourne, that it is
now regarded as an exploded hypothesis.

As a matter of course, like all zoölogists, with the
unique exception of Edward Forbes, Darwin was opposed to the theory of change of place of the continents
and oceans.   Although an evolutionist of the most
radical type when applied to animals and plants, he
was ultra-conservative and even retrograde in his views
of the permanence [1] of seas and lands.

To this day there has been nothing but chaos in
regard to the questions agitated by the new school.
All disagree on some of the most important points;
and if Darwin, according to Gray's expression, "has
turned the world of science upside down," he has
failed to give a doctrine well based and acceptable
as an indisputable truth, like the glacial theory, the
strata identified by organized fossils, the primordial
fauna, comparative anatomy, and historic and chron-
ologic geology.

As Agassiz says: "Suppose that descent of species
is proved as correct; in what are we more advanced

[1] This is a rare contradiction of all the views and opinions advanced by
Darwin, and almost incredible from a geologist who had made a journey
round the world.   Immutability of oceans and lands is a greater heresy in
the eyes of all true practical geologists than immutability of species for
Darwinians.

in our knowledge of them? Can we dispense with a study of the organism, the embryology, the exact position each species occupies in classification?" True progress in natural history does not depend on fine theories, hypotheses, and philosophy. What is wanted are new observations, new facts, new deductions well based on facts absolutely undeniable.

We shall always have a quantity of theories. We already hear of neo-Darwinism, of neo-Lamarckism; one need not be much of a prophet to predict that we shall see several other "neos" during the next century.

Cuvier and Agassiz did not believe in the descent of species and in transformism; they knew well that species vary, that intermediate forms exist, that links are constantly found; but all this did not shake their faith in the existence of species and genera. They were unwilling to go beyond what they saw with their own eyes and what they touched with their own hands.

Slow action *à la* Lyell and Darwin is very well, but principles of uniformitarianism are constantly disturbed by facts which confront every honest and careful observer, and which cannot be explained in any satisfactory way, except by the presence of paroxysms, catastrophes, and revolutions among the forces of nature as we have them now under our eyes. Before such facts, evolutionists and uniformists are at a loss, totally disagreeing among themselves. For instance, the glacial epoch has been a thorn in the flesh of Darwin, Wallace, Bates, and others, each one disagreeing with the others.

Between variations of extinct species and their affinities to each other and to living species, and transformism, or the theory of descent from common parents, there is a gap; rather, an immense abyss, over which Agassiz was not willing to leap. A bridge was needed, but natural selection seemed to him too frail a structure for so dangerous a passage. In his view, natural selection[1] was only a beautiful circumlocution; for he clearly saw that selection in natural history could not be other than natural. When naturalists like Cuvier, Agassiz, Barrande, Owen, Pictet, Lartet, Deshayes, etc., hesitate, it is simply because the conclusions presented to them are too full of obscurity and of supposition. True savants, accustomed to rigid scientific methods and exact principles, do not like to move in the dark. It is possible that for some of them,

---

[1] Darwin had obtained his idea of natural selection from the work of Malthus, as he says in his introduction to the "Origin of Species," and also in a private letter, published by Haeckel. The idea was not original with him, but only an application to all the animal and vegetable species of what Malthus held to be the principle governing the human species in its struggle for existence. Curiously enough, Wallace also got the same idea of natural selection from reading Malthus, while in camp in the jungles of Java, and he at once outlined the "revelation," which had come to him almost as an "inspiration." After all, it was simply a coincidence in the minds of two men who were interested in the same subject and were trying to theorize in regard to the progress of life through some sort of agency. Malthus furnished "natural selection," and at once both Darwin and Wallace found that their ideas had fashioned themselves into a complete system. So Malthus is the revelator and inspirer of what has been called by an enthusiastic admirer of both Darwin and Wallace "the greatest synthetical emanation of the scientific mind of our day." To this case the saying of Marshal Canrobert after the cavalry charge at Balaklava, "C'est beau, mais ce n'est pas la guerre," may be applied, with a slight change: "C'est beau, mais ce n'est pas l'origine des espèces."

like Huxley, the way may seem clear. However, even Darwin declares that it is not easy, and that he was constantly troubled by hesitation, and even doubts. Agassiz was perhaps too cautious, but no true naturalist will blame him, and the position he took in the controversy has been fully justified.

All branches of natural history, except mineralogy, are now in a transitory state, and our ignorance on many most important points is very great. In fifty years, our successors will be in a better position to form a judgment. The records of a Cuvier and of an Agassiz, with their admirable works on classification, comparative anatomy, palæontology, embryology, glacial doctrine, can await the test of time.

By way of résumé, we may say that at present the theory of descent as set forth by Lamarck and Darwin has not been established by incontestable facts and observations. Agassiz was unwilling to abandon the method of exposition of facts which he had found established in science, and to substitute in its place metaphysics and hypotheses; he clung to observation and experiment. Man has not yet found the secret of creating species; it is true, man has the power of destroying species, as he has already shown by the extermination of several species of animals. But the question of mutability of species and the method of effecting it are still reserved for future observers; and not until we possess unquestionable proof of their soundness will transformism and descent be accepted in science. Notwithstanding all that has been advanced as to pre-destined *evolution*, by Naudin, Minart, Koeliker, and

others, the bold assertions of Haeckel,[1] the natural
selection of Darwin and Wallace, are hypotheses insuf-
ficient to prove the reality of the origin and descent
of species.

De Candolle insists that transformism is no longer
an hypothesis, but a proved fact, and that the only
hypothesis lies in the explanation of the process of
variation of species and their propagation. Herein
is the whole difficulty. Agassiz has proved that each
individual, in his embryologic development, passes
through forms analogous to those of species which
have existed in geological times. If the Darwinists
can replace their hypothesis of process of variations

---

[1] Haeckel's attacks upon Agassiz's character, calling him an "hypo-
crite and a charlatan," are happily unusual in natural history. At all
events, they do not prevent him from making use of Agassiz's discoveries,
as it is proved by Alpheus Hyatt, who says: "Therefore, while the law of
correlation of the stages of development and those of the evolution of the
phylum may, if one chooses, be called a law of biogenesis, it is more
accurate to consider it a law of correlation in bioplantology; or, better
still, the law of palingenesis, or regular repetition of ancestral characters,
which very nearly expresses what the discoverer, Louis Agassiz, saw and
described. The fact that Agassiz was wrong in his theory, not believing
in evolution and not recognizing the meaning of his laws in this sense,
does not absolve those who profit by his labours from recognizing his
discovery of the facts, and his obviously full acquaintance with the law
and its applications to the explanation of the relations of organisms. It
is Agassiz's law, not Haeckel's" ("Philogeny of an Acquired Characteristic," by Alpheus Hyatt; "Proceed. Amer. Philosophical Society," Vol.
XXXII., p. 390; Philadelphia, 1894).

Ernst Haeckel is trying to play — in the origin of species — the rôle of
a Mahomet, and like him is very intolerant against all those who do not
accept his "creed" and use his method of doubt on the problem of life, as
his last work, "Monism. The Confession of Faith of a Man of Science,"
London, 1894, sufficiently proves. His preconceptions on matters which fall
within the provinces of research and discovery are anything but scientific.

of forms, by proofs based on observations easily repeated and accessible to every one who studies species, then we shall understand the origin of species by accidental transformation, which they want us to believe. The theory of the followers of Cuvier does not differ so much from that of the transformists as is generally supposed. The plan of both is the same; in both, intermediate species have always been recognized, and the discovery of links between past forms and new ones is mainly due to the researches of Cuvier, Agassiz, Owen, and others. But it is the part played by accident, as a sort of mechanical process constantly made use of by Darwin and his school, which fails to explain an enormous amount of palæontological and biological facts, which are all left to be accounted for by pure hypothesis. Suppress all hypotheses, *if possible*, and then the two schools of immutability and mutability of species will unite. But as long as hypotheses are the main factors in the problem, it will remain a problem, and not a final solution.

If natural selection or other expedients proposed by Darwin and his school will account for the origin of species, the mechanical process resorted to should not be difficult to get at. Laboratories for biological research exist now in great numbers, in both hemispheres, and if it is as simple as Darwin, Wallace, and Haeckel are inclined to think, we shall before long have new species to add to the catalogues of plants and animals. If, on the contrary, no new species is produced, we shall be obliged to have recourse, in some way or other, to the " Supreme Will and Power," according to Lyell's phraseology, or to the creation or at least sudden appearance of organized being, which Cuvier and his school have maintained as the only rational hypothesis. " Nous verrons!"

CHAPTER XIX.

1858–1864 (*continued*).

During all these years social life was at its height with Agassiz. In August, 1858, the "Saturday Club" made a summer expedition to the Adirondacks, under the leadership of the poet — afterward diplomatist — James Russell Lowell, who was the youngest and the most energetic of the "philosophers' camp." A roughly built shelter, with a roof of fir bark, on the shore of Follansbee Pond, a small lake in the Raquette Mountains, not far from Keeseville on Lake Champlain, served as tent for the whole party, which was composed of Agassiz, Ralph Waldo Emerson, Lowell, John Holmes, Dr. E. Howe, Judge Hoar, A. Binney, Jeffries Wyman, and a few others. The life was rather rough; all were in flannel shirts, red or blue, and slept wrapped

up in blankets on fir boughs.   Game and fish were
abundant, and the fare good.   Longfellow had declined
point blank to go, because Ralph Waldo Emerson,
whose mind was always wandering among the temples
of Greece and Rome, and seldom concerned with reality,
had taken a gun.   When asked why he refused to join
them, Longfellow's ready answer was, because "some-
body will be shot."   However, nothing was shot, except
deer.

It was the custom at the "philosophers' camp" every
morning after breakfast to practise firing at a mark,
during which time Agassiz, Wyman, and an assistant
would dissect and prepare their specimens.   One day
some one asked Agassiz to shoot at the mark; and on
his hesitating to accept the offer of a rifle, the whole
company joined in the request, urging that a man with
such eyes must be a capital shot.   Not one of them
knew or imagined that Agassiz had never fired a shot
in his life.   Finally Agassiz took aim, fired, and put
the ball in the eye of the mark, which gave occasion
for much applause and many compliments.   Further
solicitations found him immovable, and he firmly declined to fire another shot.   He had been too lucky to
try again, and this was actually the only shot he ever
fired.[1]

[1] In a letter to Cuvier, written about 1827, Agassiz mentions that he
"practised arms, the bayonet and sabre exercise" (Mrs. Agassiz's "Life
and Correspondence of Louis Agassiz," Vol. I., p. 108), which seems to
indicate that he then used a gun.   But it was not so.   William Schimper,
the brother of Karl, when he joined Agassiz at Munich, had just left the
Baden military service as a non-commissioned officer; and he drilled
Agassiz in the use of a gun, without firing it.   Agassiz always declined to
fire a shot, and would give no reason for his refusal.

In Boston and Cambridge dinner parties succeeded each other so rapidly that it was a wonder that any man could stand such a strain on his digestive powers. Agassiz was a member of all the fashionable clubs of the time, and besides was a welcome guest at the hospitable tables of all the leading families of Boston. A description of one of these meetings will suffice. Dr. Holmes, the great humourist and poet, says, " At the other end of the table [of the " Saturday Club "] sat Agassiz, robust, sanguine, animated, full of talk, boylike in his laughter." They lingered long round the table, while hour after hour passed in animated conversation, in which *bon mots* and repartee were exchanged as rapidly as a discharge of fireworks — an encounter of anecdote, wit, and erudition. At such times Agassiz was at his best, with his inexhaustible *bonhomie*. With a lighted cigar in each hand, he would force the attention of every one around him. Excited by the pyrotechnic wit of James Russell Lowell, Judge Rockwell Hoar, and the author of the " Autocrat of the Breakfast Table," Agassiz, whose vivid imagination was always on the *qui-vive*, was not a man to let others eclipse him. Then would come one of his made-up stories — a mixture of dream and science. He knew perfectly well that it was a fiction, and the first time he told it he hesitated a little. If he thought any one in the company was doubting its truth, he would look at him with a dumb request not to betray him. On the next occasion he would repeat the same story without any hesitation ; and the third time he told it, he was sure that it had really happened, and was true. Agassiz would have been very truthful, if he had had less fire and brilliancy in his imagination, always too easily excited. In principle he was honest, because he believed all that he said. For him the Italian proverb, " si non e vero e ben trovato," was an article of the code of conversation in after-dinner talk among witty gentlemen. Very appreciative of a well-served table, of witty conversation, and of the company of ladies, his *gaulois* spirit formed a relish, as it were, for his more serious and guarded American friends. As Lowell says of him, " Blood runs quick in his veins, and he has the joy of animal vigour to a degree rare among men — a true male, in all its meaning."

Lowell was a special favourite with Agassiz, and knew him thoroughly. As Agassiz was always a great walker, he and Lowell, after a long sitting at the " Saturday Club," in the early hours of morning, would come back to Cambridge on foot, Agassiz continuing his confidential *gauloiseries*, begun " under the rose," until the joyful companions were forced to separate, a parting which Lowell has so well described in his memorial poem entitled " Agassiz " : —

> " At last arrived at where our paths divide,
> ' Good night ! ' and, ere the distance grew too wide,
> ' Good night ! ' again ; and now with cheated ear,
> I half hear his who mine shall never hear."

While returning from one of these Saturday Club meetings, Agassiz was greatly shocked by the sudden illness of his friend, President Felton, who suffered a severe attack of heart disease, as they were walking home. It proved a first warning of death ; and some months later, on the 26th of February, 1862, a second attack prematurely ended the life of one who had been a most devoted friend and true brother to Agassiz.

Ever since his arrival in Boston, in 1846, Agassiz had been not only a welcome guest, but a great favourite, at the house of Mrs. George Ticknor. Mrs. Ticknor's literary "salon" exerted, during the middle of the nineteenth century, a great influence on New England society. There gentlemen and ladies, distinguished for their literary attainments, their education, their high official position, met daily. Foreigners, as well as Americans, came and went constantly under this charmingly hospitable roof; and Agassiz, when in Cambridge, was one of the most assiduous "habitués." Although not scientific, both Mr. and Mrs. Ticknor enjoyed the conversation of savants. Sir Charles Lyell and Lady Lyell were guests at the Ticknors' during each of their visits to America; and Agassiz ran in almost every time he came to Boston, sure to find there, not only friends, but sympathizers and often helpers of his never-ending schemes for the progress of natural history in North America.

Another house where Agassiz was often a most welcome guest was that of the director of the Perkins Asylum for the Blind at South Boston. Dr. S. G. Howe, the philanthropist, and his gifted wife, Mrs. Julia Ward Howe, appreciated Agassiz at his real value as soon as they knew him. The friendship was reciprocated; and it was no small privilege and enjoyment to hear a conversation between him and Mrs. Howe, both geniuses, and spirited and witty to a rare degree.

Charles Sumner, although a good friend, was too much engrossed by politics for Agassiz, who never much relished political societies and meetings. Natural history discussions left no time for other debates. Outside of his natural history pursuit, pictures, especially landscapes, were the only things which attracted him, although he had little time to devote to them. He saw at once the quality of a picture; and I have seen him lost in admiration before Alpine landscapes by Calame, Diday, and Töpffer, or beautiful "paysages" of the Jura Mountains by Gustave Courbet.

At the beginning of the Civil War Agassiz received, through M. Jules Souchard, the French consul at Boston and his personal friend, a message written by order of the Emperor Napoleon, asking information in regard to the acclimatization of several marine animals living in the American Atlantic. With his usual promptness to help in anything relating to natural history, Agassiz took the trouble to send, in charge of his old friend Burkhardt, a whole cargo of living *Mya arenaria, Venus mercenaria, Pecten concentricus, Homarus americanus, Mactra solidissima,* and *Mytilus edulis,* which might be experimented with at the oyster station near Havre. The passage from Boston to Liverpool, which was made just at the equinoctial time, at the end of September, 1861, was very stormy and long; and almost all the animals died during the voyage, notwithstanding the care of the captain of the steamer, James (afterward Sir James) Anderson, a special friend and great admirer of Agassiz.

Agassiz, Burkhardt, and Anderson took every pre-

caution possible in this first experiment of sending living marine animals across the Atlantic; but owing to
adverse circumstances, not only the long and protracted
journey over the Atlantic, but also the journey from
Liverpool to Havre, only two hundred live specimens
of *Venus mercenaria* arrived at La Houge de Saint-
Waast, on the island of Tatihou, in the marine aquarium,
near Normandy's coast, Havre.

Burkhardt was presented to Napoleon at the Tuile-
ries, who thanked him for the part he had taken in the
difficult task of conveying such an unusual cargo; and
Agassiz received through his friend Souchard the imperial thanks and the cross of officer of the Legion
d'honneur. The failure did not discourage further
experiments. On the contrary, the difficulties encoun-
tered by Burkhardt were taken into consideration; and
other cargoes were sent across in 1862, with complete
success, by a lieutenant of the French navy, M. Philippe
de Broca, sent for that special purpose by the Secretary
of the Navy. M. de Broca carefully followed all the
instructions and advice given to him by Agassiz; and
although two-thirds of the animals sent over at different
times during his stay in the United States died on the
voyage, a third of them, about ten thousand specimens,
arrived alive and in tolerably good condition.

Agassiz much enjoyed the society of Jules Souchard,
the French consul, and he and Agassiz arranged a
weekly meeting half-way between their two houses.
Every Sunday afternoon each set forth from his home
at the same hour, and walked toward the other until
they met. Then they continued the walk together as

far as the house of one or the other, alternating each
week, whence, after enjoying a glass of French wine
and cigars, the visitor would return by a horse car.
This friendly arrangement lasted several years, until
Souchard returned to France in 1867.

The Civil War was a terrible hindrance to the progress and prosperity of Agassiz's Museum. First of all,
almost half of his assistants and pupils left to enlist
in the army. Three of them died there, — N. Bowditch,
C. A. Shurtleff, and A. P. Cragin. Nathaniel Bowditch, son of Dr. Henry Bowditch, was killed on one
of the battle-fields of Virginia. Albert Ordway rose
rapidly to be colonel, and finally brigadier-general, of
United States volunteers. He had had charge of the
crustacea at the Museum, and had begun good work on
the trilobites, but he unfortunately never returned to
the Museum, and has been lost to science. Alpheus
Hyatt remained over a year in Louisiana on the staff
of General Banks, the commanding officer, and after
good service returned as captain of volunteers in 1863,
several months after his time of service was over, for
he was so appreciated by his chief that General Banks
would not allow him to return earlier. Hyatt, more
than any other assistant of the Museum, deserves credit
for having enlisted for active service, because, educated
in Maryland, his family were in sympathy with the
South, and in enlisting in the Northern army, he en-
countered their strong opposition.

Albert Bickmore deserves more than a passing
notice, for he made use of his spare time in the army
as a nine-months soldier to make a splendid collection

of the marine animals, more specially shells, of the coast of North Carolina. Being only a private in an infantry regiment, he was detailed upon special hospital duty near the seashore. There almost daily he explored the beaches, going often so far as to trespass on the Confederate lines, wandering in his pursuit of shells, on rather dangerous ground for a boy in blue, for he always wore his military dress. The least that could have happened to him was to be made a prisoner of war, but somehow his pursuit was so earnest, he seemed so indifferent to danger, so absorbed in collecting shells, that the Confederates looked on him as a sort of inoffensive "crank," and let him alone. Bickmore is now the efficient curator of the American Museum of Natural History at the Central Park, New York City. He was a modest pupil, and does credit to Agassiz.

Theodore Lyman, appointed colonel on the staff of General Meade, served with distinction until the end of the war.

The other pupils and assistants did not enter the army. Mr. Alexander Agassiz was confined to the halls and laboratories of the Museum, where he did excellent and valuable service.

I was the only person attached to the Museum who crossed the belligerent lines. In the fall of 1863, I ventured on a geological exploration in Nebraska. At that time there was no railway in Iowa, and to go from Burlington, Iowa, to Omaha, I was obliged to pass through all the northern part of Missouri, the most rebellious part of that state. Railroad trains were constantly held up by guerillas and bushwhackers on the

Hannibal and St. Joseph line, notwithstanding the gar-
risons of United States troops stationed in and around
block-houses at several important points of the road,
and the day before I passed, the train had been fired at
and stopped by a party of guerillas. However, I arrived
safely, and embarked at St. Joseph to go up the Missouri River. My exploration was very successful, and
I brought back to the Museum important collections of
fossils. Agassiz congratulated me on my return, saying
that he was not without some apprehension as to my
safety.

Several other pupils of Agassiz, who came during
1862 and 1863, have since distinguished themselves in
natural history. I would mention William H. Dall, the
efficient curator of invertebrates at the United States
National Museum; William H. Niles, professor of geology and geography at the Massachusetts Institute of
Technology; Horace Mann, the explorer of the Sandwich Islands; and the entomologist, P. R. Uhler.

As I have said, the Civil War retarded the progress of
the Museum, for it was no time to ask for money when
all the resources of the country were required to carry
on the war. Still, in 1863 Agassiz obtained from the
Legislature of Massachusetts a grant of ten thousand
dollars for the publication of an illustrated catalogue of
the Museum — the best proof of his great popularity
among the inhabitants of Massachusetts.

The want of money became so pressing in 1863, that
Agassiz bravely made a last effort to obtain it, by a
grand lecturing tour in the West. The Museum had
existed for only three years then, and to surmount the

scarcity of funds, increased by the depreciation of government money, was of prime importance almost for the very existence of the new institution. With his ordinary pluck and courage Agassiz did not hesitate for a moment, and plunged into the scheme without a doubt of his success, and at the same time trusting to the herculean strength of his constitution to bear the strain. However, he presumed too much on his endurance, and those near him realized his danger. Although everything which love could imagine was done to help him and spare unnecessary fatigue, Agassiz returned from his tour, on which his admirable wife had been always at his side, much exhausted and broken down. At the age of fifty-six the strain was too great. Going from town to town, from Buffalo, Cleveland, Detroit, Chicago, to St. Louis, lecturing always before crowded audiences, from December, 1863, until March, 1864, was a veritable *tour de force*, which it would have been absolutely impossible for him to repeat. Happily, although money was always wanted in larger quantities than it came, Agassiz was not again embarrassed by lack of pecuniary aid. To be sure, he was always in pursuit of money, pressing the Massachusetts Legislature not to forget his museum, but aid came more readily from private individuals, and the periodical crisis concerning money, to which he had been subjected all his life, at last passed away, never to return.

When in Chicago he planned an excursion to Burlington, Iowa, on his way to St. Louis. In my visit to Burlington in September, 1863, I had been absolutely astounded by the extraordinary wealth of fossil crinoids displayed in three private collections belonging to citizens of that town. After my return, I had spoken of them in such glowing terms to Agassiz, who knew how difficult it was for me to become enthusiastic over anything touching collections, that he resolved to see for himself. He found that the reality surpassed all his expectations, and at once he was eager to have all three of these collections transferred to his museum. The temptation proved too strong to resist, and he purchased one, promising as soon as his means would allow to purchase the others, which he did a few years later. Finally, the great and unique collection of Dr. Charles Wachsmuth, containing almost all the typical specimens described of Western fossil crinoids, was safely stored in the numerous drawers of the Agassiz Museum, under the direction of Dr. Wachsmuth himself. What a devourer of collections Agassiz was!

At St. Louis Agassiz enjoyed the society and companionship of Dr. George Engelmann, an old classmate at the Heidelberg University, and their reminiscences of student life and intimate association with Karl Schimper and Alexander Braun, revived their young days, when morphology of plants was the constant subject of their thought and talk. In fact, Engelmann's first paper, written as his thesis, or inaugural dissertation, for his degree of Doctor of Medicine, in 1831, was entitled "The Morphological Monstrosities of Plants." Engelmann's character and spirit were most congenial to Agassiz, who appreciated him highly, for, in the words of his life-long friend, Dr. A. Wislizenus, Engelmann "was firm and decided. He did not rely upon speculations in his scientific researches, but on facts only, ascertained by severe and searching studies. He was strictly true in scientific matters." Both have followed constantly the same principles in their researches and classifications, and Agassiz much admired Engelmann's great works on the *Cactaceæ*, the *Yucca*, the *Agave, Jaucus, Coniferæ*, the *American oaks*, etc., etc. After the passionate discussions on the "Origin of Species," it was a great comfort to Agassiz to find a botanist not given to speculations and theories, but standing firmly on plain and proved facts.

Agassiz passed the summer at Nahant, where he had a seaside laboratory close by the cottage of Mrs. Agassiz, a charming place of resort for a naturalist, much enjoyed by him and his children. Naturally Agassiz kept every one round him busy, directing the microscopical studies and researches, superintending the drawings, and giving his leisure time to dreams of schemes to increase the usefulness and wealth of his dear museum.

Now a word in regard to the name given to his creation, no longer in its infancy, but in full vigour. *Museum of Comparative Zoölogy* is a very long title, especially for common people; and for a man like Agassiz, who always courted popularity, it may seem strange that he made such a choice. First, the name is a pleonasm, for it is impossible to work at zoölogy without making comparison. It is not so with anatomy; and Cuvier, in creating *comparative anatomy*, used a proper term for a new science. Agassiz knew

this perfectly well, and one day, in a mood of confidence with me (after his forty-fifth year Agassiz, until
that time extremely frank, saying even more than was
prudent, became rather reserved and reticent), admitted the fact that the title was not very appropriate.
"I did not want the name of any patron or benefactor
given to it," he said; "Mr. Peabody, the generous
American banker of London, has informed me that
he will endow the Museum with a large sum of money,
but on the condition that it shall bear his name; that
I cannot accede to." "Of course," said I, looking
him full in the eyes, "it will be *Agassiz's Museum.*"
"Yes," he answered feebly, "you have ferreted my
secret."

To repeat this now is not to betray my illustrious
friend. Every one, savant or illiterate, native or foreigner, calls it "Agassiz's Museum," notwithstanding
the great sign, "Museum of Comparative Zoölogy,"
sculptured in big letters over the gate. It is simple
justice, and the reward conferred by universal consent
on the man of genius who created it, from nothing,
with his brain and invincible will. After all, men are
not ungrateful, even in a republic.

His summer at Nahant not having given the relief
that he expected, as a means to restore his health and
get more strength for the next year's work at the
University, he made, during the whole of September,
1864, an extended excursion into Maine, looking for
glacial remains. He extended his researches for moraines and oesars—now called drumlins—from Bangor
and Katahdin to Mount Desert, and carefully studied

what are called "horse-backs." His Alpine experiences
of twenty and more years before came vividly before
him, when in presence of the morainic material accumulated all over the state of Maine, and after returning
home he dictated one of the best articles he contributed
to the "Atlantic Monthly," under the title, "Glacial
Phenomena in Maine."

CHAPTER XX.

1865–1867.

EVER since 1828, when a student at Munich, he
undertook the publication of the "Brazilian Fishes"
collected by Spix, Agassiz had cherished the hope of
some day exploring the basin of the Amazons and seeing Brazil. It may almost be called his hobby; and
his relations with the Emperor of Brazil, with whom he
had exchanged letters on scientific subjects, so much
increased his desire that he resolved at the beginning
of 1865 to carry out his plan of a visit to Rio Janeiro.
His health had been gradually giving way ever since
his illness at Charleston in 1853; and the attack he
suffered at Cambridge in 1863 made it important for
him to seek a change of scene and climate, with rest
from work. Brazil was his lifelong desire, and towards

it he bent all his energy. Mr. Nathaniel Thayer, a friend, and at the same time one of the richest men of Boston, whom Agassiz had succeeded in enlisting as treasurer of the board of trustees of the Museum of Comparative Zoölogy, provided most liberally for six assistants and all the expense of collecting and forwarding the specimens to the Museum.

Agassiz embarked at New York, on the 2d of April, 1865, and arrived at Rio Janeiro the 22d of the same month. He was accompanied by his wife, Burkhardt as artist, J. G. Anthony as conchologist, Frederick C. Hartt and Orestes St. John as geologists, J. A. Allen as ornithologist, and a preparator. Besides these, six volunteer assistants, among them Mr. William James, had joined the expedition. The journey lasted sixteen months, ten of which were passed on the Amazons. Two of the assistants, Anthony and Allen, were soon compelled by poor health to leave for home. The artist, Burkhardt, although a constant sufferer during the whole trip, bravely continued his work until the end, drawing living fishes in their natural colours. But the exertion was too much for him ; and this faithful companion of Agassiz returned home with such impaired health that after ten months of sickness at Cambridge, he died in the house of Mrs. Pauline Shaw, née Agassiz, whose kind heart and grateful remembrance of many kindnesses bestowed on her by Burkhardt when a child and a young lady drew her to the sickbed of the old man, whence she took him in her carriage to her beautiful home in Jamaica Plain, where Burkhardt breathed his last, after a few days of the most affectionate nursing and tender attention from his hostess.

Dom Pedro Secundo, Emperor of Brazil, received Agassiz in the kindest and most liberal and generous manner. It was a great pleasure for him, a scientific *dilettante,* to receive such a naturalist under his roof and in his empire. Educated partly in Switzerland, Dom Pedro had heard of Agassiz's researches on the glaciers and on fishes; and as his turn of mind was decidedly scientific, he had read more of Agassiz's works than any one else in Brazil. From their first meeting the two men were friends. His Majesty enjoyed Agassiz's immense store of knowledge, his brilliant spirit, and the charm of his conversation, while Agassiz, on his part, was rather surprised to find a crowned head so well instructed in geology, the glacial theory, and other scientific questions. Dom Pedro rendered every possible aid to the expedition; and from the time Agassiz put his foot on Brazilian soil until he left, the Emperor showed his great interest in the success and comfort of Agassiz and his party, and in these respects he succeeded admirably.

As soon as he arrived, Agassiz divided his party into several smaller ones, some to go to the interior, others to explore the coast; and as usual with him, and at the request of the Emperor, he delivered a course of lectures, open to all, without charge, at the Collegio Dom Pedro II, before a very large audience of gentlemen and ladies, headed by the Emperor with his whole family. Agassiz spoke in French; and it was a great pleasure to him to address an audience in his native tongue, after so many years of constant lecturing in the English language, of which he was never a complete master. After a visit by railroad to the province of Minas-Geraes, Agassiz sailed from Rio, the 25th of July, for his Amazonian journey. The Emperor had detailed Major Joâo M. da Silva Coutinho of the engineer corps ˙to accompany Agassiz during his whole exploration, — an admirable selection; for no Brazilian was better acquainted with the region of the Amazons River than he, having been engaged there for several years in scientific surveys. " His assistance was invaluable to us throughout the journey," says Agassiz; "and he became my intimate associate in all my scientific undertakings in Brazil. During eleven months of the most intimate companionship I had daily cause to be grateful for the chance which had thrown us together. I found in Major Coutinho an able collaborator, untiring in his activity and devotion to scientific aims, an admirable guide, and a friend whose regard I trust I shall ever retain."

The attachment was reciprocated by Coutinho, who became a great admirer of Agassiz. A Brazilian by birth and education, he was very fond of the wild Indian life of the tropical forests, and acted during all Agassiz's journey on the Amazons as an Indian scout looking after natural history specimens. In one of his rambles he put his foot on a big boa-constrictor, taking it for an old log fallen across the path. An unusual noise above his head put him on his guard and notified him of his mistake, but he did not choose to follow the beast into the swampy thicket. Coutinho was one of the most progressive men in Brazil. He was one of the promoters of the construction of railroads in all parts of the country, and distinguished himself greatly as a government engineer. I became very strongly attached to Coutinho afterward at Paris in 1867, 1868, and 1877, and appreciated highly the charm of his society as a geologist, geographer, and friend. He died in Paris, October 11, 1890, while yet in middle life, from an illness contracted in the unhealthy wilderness of the Brazilian seacoasts, during his railroad surveys.

In the following paragraph I quote Agassiz's words in regard to his expedition : —

Once in the waters of the great river (the Amazons), I divided my forces, in order to survey simultaneously various parts of this vast fresh-water system, wishing to ascertain how far the distribution of its inhabitants was local, or whether the same species might be found at the same moment in different parts of the main stream and its tributaries. This precaution led to results which amazed me, though I was in part prepared for it by my knowledge of their aquatic fauna. Not only did I find the number of species in these waters exceeding by thousands all former estimates, but I found their location so precise and definite, that new combinations occurred at given intervals along the main stream, while every forest lake and all the lesser watercourses had their special fauna. I neglected no opportunity of verifying the accuracy of my results, visiting the same regions at different seasons of the year, repeating my collections that I might have the fullest means of comparison, and, as I have said, stationing my parties at considerable distances, in order that by making simultaneous collections, we should ascertain what was the range of the species (" Special Report of the Directors of the Museum of Comparative Zoölogy," in " Annual Report for 1866," p. 14; Boston, 1867).

The Amazonian Steamship Company placed a fine steamer, furnished with everything needed by the whole party, at Agassiz's disposition for one month, while later a ship of war was sent up by order of the Emperor for the use of Agassiz during the remainder of his stay in the waters of the Amazons, to replace the Company's steamer, and wherever Agassiz arrived he found that directions had been given to the local authorities to furnish him with canoes, men, and whatever else he might require for his scientific researches.

The first station on the Amazons was Pará, then Manaos, Tabatinga, and Teffé.   An excursion was made on the Rio Negro as far as Pedreira, where the stones in the bed of the river were so numerous and large, that the channel was too dangerous for the war-steamer *Ibicuhy*.

To Agassiz, as well as to Bates, Wallace, and Martius, the valley of the Amazons seemed the paradise of naturalists. His enthusiasm and admiration of everything he met knew no bounds, while he busied himself in collecting animals of all sorts, plants, more especially all the palms and ethnological specimens, and observed with his keen and searching eyes everything from men to insects. He filled to its utmost capacity with specimens the war-steamer, even the deck being encumbered with trunks of palm trees. For him the basin of the Amazons was a "fresh-water ocean," with an archipelago of islands. The character of its fauna is also oceanic, and its most noticeable feature is the abundance of cetaceans through its whole extent.

The health of Agassiz during his stay on the Ama-

zons was good, but it is no wonder that at last he was
overcome by fatigue, and though not actually ill,
was exhausted by incessant work, and by the contem-
plation, each day more vivid and impressive, of the
grandeur and beauty of tropical nature. On his return to Pará, where he arrived in February, 1866, he
was so tired as to be unable for several days to exert
himself even to write letters. The climate had affected
and enervated him more than he had at first thought.
If Agassiz had made an exploration of the Amazons
thirty years before, when still in the prime of life, the
results would have been very different. Although he
had undoubtedly acquired great practical experience on
all zoölogical questions, and made good use of that
experience during his journey up and down the great
Amazons, he was too old to make a full use of his rare
ability as an observer. His mind was no longer so
elastic as at the time of his sojourn on the glacier of
the Aar, and although his eagerness to collect specimens was as great as ever, he had no longer the bodily
strength to make full use of them.

Certainly his " Brazilian expedition, fitted out and
sustained by individual generosity, was treated as a
national undertaking, and welcomed by a national hospitality," and Agassiz succeeded in bringing safely
home to his museum the treasures he accumulated in
Brazil; but it remained to work them up, to classify
them systematically, and, as he himself says, " a critical
examination of more than eighty thousand specimens
cannot be made in less than several years." Unhappily this critical examination, for some reason or other,

was not only never finished, but was hardly begun, except some years later in the case of the fishes only, by his assistant, Dr. Steindachner. This is much to be regretted; for if he had accomplished what he planned before he started on that expedition, it would have resulted in a great advance in our knowledge of the geographical distribution of aquatic animals. In his own words, " One of my principal objects during the whole journey was to secure accurate information concerning the geographical distribution of the aquatic animals throughout the regions we visited. Upon this subject we had little precise knowledge, — even the best known among the fishes, reptiles, etc., of the Brazilian waters being entered in our zoölogical records simply as living in Brazil, or more generally still, as found in South America. As the distribution of species lies at the very foundation of the question of their origin, I have aimed at ascertaining as far as possible what are the areas and limits of their localization." It is a pity that he did not accomplish this localization!

Before leaving Pará, Agassiz delivered a lecture on the physical history of the valley of the Amazons, which was afterwards published in " The Atlantic Monthly," Vol. XVIII., July, August, 1866, pp. 49–60, 159–169, and reprinted in " A Journey in Brazil," Boston, 1868. The month of April, 1866, was devoted to an exploration of the province of Ceará, with the special purpose of looking for traces of ancient glaciers. The time was not propitious; it was the rainy season, and it was not easy to reach the Sierra of Aratanha in the interior of the province. However, Agassiz succeeded in finding the glacial phenomena as legible as in any of the valleys of Maine, or even the Cumberland Mountains of England, with medial, lateral, and frontal moraines, at a level of only eight hundred feet above the sea in latitude four degrees south.

The roads in Ceará, during the rainy season, are so bad that the only way to travel is on horseback. This was the only time that Agassiz ever rode, and it was so trying and disagreeable to him, that he made most of the journey, especially the mountain scramble, on foot, notwithstanding the mud and the consequent pitching, tumbling, and sliding. He never repeated the experience, and nothing could have induced him then to mount a horse but his great desire to see moraines under the tropics.

By the end of April Agassiz was back again at Rio Janeiro. During the month of May he delivered at the Collegio de Pedro II, a series of six lectures, which was attended by the *élite* of society, ladies as well as gentlemen. It seems that before Agassiz came to Brazil ladies did not make their appearance at public lectures. It was certainly a progress in Brazilian customs that senhoras were allowed to follow a course of lectures, and Agassiz was much pleased with the sympathetic reception given by his Brazilian audiences. The lectures, delivered in French, were stenographically reported, were then translated into Portuguese by the French naturalist, M. Felix Vogeli, and published under his direction at Rio, bearing the title, "Conversações Scientificas sobre o Amazonas," and had a large circulation.

A last excursion in the vicinity of Rio to the Organ Mountains during June terminated Agassiz's journey in Brazil. He embarked the second of July for the United States, and arrived in Cambridge in August, 1866.

The main results of the Brazil expedition were, first, the great collections of all sorts, which were stored in the Agassiz Museum, awaiting final arrangement; second, two series of lectures at the Lowell Institute, in Boston, and at the Cooper Institute in New York, on the scientific generalizations growing out of the expedition; third, a volume of five hundred pages, entitled, "A Journey in Brazil, by Professor and Mrs. Louis Agassiz," Boston, 1868, with a French translation by F. Vogeli,[1] Paris, 1869; and fourth, a few articles published in reviews, and in the "Bulletin Société Géologique de France" — these last on the geology and with the collaboration of Coutinho and the present writer.[2]

The "Journey in Brazil" was a disappointment to the public in general, and more especially to the naturalists and personal friends of Agassiz. "The Naturalist on the River Amazons," by H. W. Bates, a work of genius, had somewhat spoiled the Amazons as a field of research, and had led people to expect more important

---

[1] The French edition is more complete and more valuable than the English, containing three times as many illustrations and maps and also some additional notes.

[2] "Lettre de M. Agassiz à M. Marcou sur la géologie de la vallée de l'Amazone, avec des remarques de M. Jules Marcou" ("Bull. Soc. Géol. France," Vol. XXIV., p. 12, December, 1866). "Sur la géologie de l'Amazone," par Agassiz et Coutinho, with notes by Marcou ("Bull. Soc. Géol. France," Vol. XXV., p. 685, May, 1868).

results from a naturalist of such repute as Agassiz.   It
is true that Bates remained there eleven years, while
Agassiz passed only ten months, but the store of
knowledge possessed by both was so different, that it
was natural to expect not only something startling, but
also something which might have some effect on the
theory of the origin of species, in a Cuvierian way.
The only part which can be called Cuvierian is an
appendix on "The Permanence of Characteristics in
Different Human Species."

In Agassiz's volume the personal adventures and
incidents of travel are rather tame and the style dull
and heavy, not in harmony with the usual brilliancy and
spirit of the great naturalist, while in Bates's volume
the narrative is most attractive, whether he speaks of
adventures, incidents, or purely scientific matter, and
the style full of animation.   The difference is due
mainly to their mode of travelling, one journeyed in
state, as it were, while the other, alone, and with very
scanty, sometimes without any pecuniary means at his
disposal, forced his way with great difficulty.   Besides,
Bates was there in his prime, and wrote his volume
himself.

However, Agassiz's influence on the progress of
natural history in Brazil was very great, so far as anything makes a lasting impression upon a population
inhabiting such a warm climate; for we must not forget the *dolce far niente* of the inhabitants of tropical
regions.

It is interesting to see how Agassiz was influenced
by what he observed during his Brazilian journey, in

regard to the origin of species, for he visited exactly
the same ground that Henry W.. Bates and Alfred R.
Wallace had visited a few years before. In a letter to
his old friend the ichthyologist, Sir Philip de Grey
Egerton, dated Cambridge, March 26, 1867, Agassiz
says, " I have about eighteen hundred new species of
fishes from the basin of the Amazons! . . . It sug-
gests at once the idea that either the other rivers
of the world have been very indifferently explored, or
that tropical America nourishes a variety of animals
unknown to other regions. . . . My recent studies
have made me more adverse than ever to the new
scientific doctrines which are flourishing now in England. This sensational zeal reminds me of what I
experienced as a young man in Germany, when the
physio-philosophy of Oken had invaded every centre of
scientific activity, and yet, what is there left of it ? I
trust to outlive this mania also. As usual, I do not ask
beforehand what you think of it, and I may have put
my hand into a hornets' nest, but you know your old
friend Agassiz, and will forgive him if he hits a tender
spot." [1]

Until the end of his life Agassiz considered "the
transmutation theory as a scientific mistake, untrue in
its facts, unscientific in its method, and mischievous in
its tendency" ("On the Origin of Species" in "American Journal of Science and Arts," Vol. XXX., July,
1860, p. 15).

It is just to say that the sixteen months spent in
Brazil were among the most happy of Agassiz's life.

[1] "Louis Agassiz," by Mrs. E. C. Agassiz, Vol. II., pp. 646–647.

He enjoyed everything immensely, was never sick, only
now and then tired through over-exertion and excitement, was received with open arms by every one, from
the Emperor to planters, traders, etc., and for the first
time in his life was unembarrassed financially, being
amply supplied with money by Mr. Nathaniel Thayer.
Besides this, his wife at his side always took upon herself
a great part of the management, and shared his comfort
and fatigues, and was the recipient of many complimentary attentions on the part of the Brazilian population.
Altogether it was a sort of triumphal scientific exploration, certainly merited after the many years of hard
work in Europe and in America of one who concentrated in himself the careful and original studies of
almost half a century.   No naturalist more deserved
such a reception than Louis Agassiz.

   General incidents in Agassiz's life, which happened
before his journey to Brazil, have been passed over, in
order not to break the narration of more important
events.   Among these we may mention the receipt of
the Copley Medal, awarded to him in December, 1861,
by the Royal Society of London, — an honour which
pleased Agassiz much, and which is considered by the
English savants as the highest reward they can bestow
on a foreigner or a native.   It was certainly well placed
this time, for few of its recipients have done so much
for the progress of natural history in both hemispheres.

   Not long after the outbreak of the Civil War, in 1861,
in its darkest hour, Agassiz took out naturalization
papers, to show his sympathy with the Union.   Until
this time he had kept his Swiss nationality, notwith-

standing his acceptance of various positions and offices in America. He felt indignant at the action of England and France in recognizing the Southern Confederacy, and did his best to open the eyes of certain European officials in this country to the right side of the question and the final results.

In March, 1863, during a session of the Board of Regents of the Smithsonian Institution, he joined Professor Bache in his scheme for the foundation of a National Academy of Science. Bache was a rather ambitious man, full of academic distinctions, and a lover of power. In 1860 Agassiz had him elected a corresponding member of the Academy of Science of the Institute of France, and from that moment Bache worked at the creation of a National Academy, to bear some analogy to the French one. Under the pretext that the government in Washington might be in want of advice, directions, and reports on scientific subjects, Bache, supported by Agassiz and Joseph Henry, obtained, through Henry Wilson, then Vice-President of the United States, an act by the Thirty-seventh Congress "to incorporate the National Academy of Science."

Agassiz, who knew the defects of close corporations with government privileges, like the Institute of France, hesitated in following Bache, as did Joseph Henry. But both had been in such intimate relationship with Bache, and the American Association for the Advancement of Science, founded in 1848, had given such scanty results, notwithstanding the influence exerted on the committee by Professor Bache and his friends, that they thought a trial might be made. Agassiz may be called one of the founders, but not the "prime mover." Returning from Washington, after the act was passed by Congress, Agassiz was certainly not an enthusiast on the subject, and even showed a dislike to talk about it, simply saying that "the National Academy was mainly to satisfy Bache's ambition for control." A friend told him that it would soon fall into the hands of politico-savants, which he admitted might be true; and, in fact, a few years after the death of Bache, Agassiz, and Henry, the National Academy of Science became, as predicted, a tool in the hands of ambitious government employees at Washington.

In 1864, before Agassiz's journey to Brazil, Dr. Brown-Séquard had taken up his residence at Cambridge, as professor in the Medical School of Harvard; and a friendship based on mutual admiration soon sprang up between them. After his return the relations became very close, and Agassiz urged Brown-Séquard to go to Paris as the best place to prosecute his physiological researches. He gave him very strong recommendations, advising the Minister of Public Instruction to create a special professorship at the Medical School. Armed with these letters of Agassiz, Dr. Brown-Séquard went to Paris at the end of 1867, and was soon appointed professor of physiology in the medical faculty, a new chair created in his favour; and later he succeeded the great physiologist, Claude Bernard, at the "Collége de France" and at the Academy of Science of the French National Institute.

Some of Agassiz's pupils have already been referred

to; others, added from time to time, require special mention. Among the assistants whom he took to Brazil were Hartt, Saint John, and William James, all three noteworthy naturalists. Charles Frederick Hartt came to Cambridge in 1862, for instruction at the Agassiz Museum. He was already in advance of all the other students as a practical geologist, having worked steadily and intelligently at the geology of a part of Nova Scotia and New Brunswick. I at once saw his value, and followed his work with interest. Although a member of Agassiz's party in Brazil, he did little in the geological field, because the special part of Brazil assigned to him was devoid of fossil remains; but he conceived a desire to see more of the country, and returned to it in 1867. This time he had greater success; and about Bahia and Sergipe he collected many fossils, and really began his geological survey of Brazil. In 1868, having been appointed professor of geology in the new Cornell University, on the recommendation of Agassiz, he organized a third expedition to Brazil in 1871; and in company with some of his pupils, he explored the Amazons River region. On this occasion he succeeded in discovering the Devonian system at Monte Alegre and Sierra Ereré, and extending the area of the Cretaceous rocks between Pará and Pernambuco. In these two expeditions Hartt showed his great capacity as an observer and a leader. After the death of Agassiz he submitted to the Brazilian government a plan for a systematic geological survey of Brazil, which was accepted. Having been consulted, I did not hesitate to recommend the scheme to the

Emperor Dom Pedro; and in 1877, in Paris, the Emperor thanked me, saying how highly he appreciated Hartt and his assistants, Orville A. Derby, R. Rathbun, and J. Branner, "all very able and conscientious young savants," as he expressed it. Unhappily yellow fever killed Hartt in March, 1878, after only two years of work on the geological survey of Brazil, which was discontinued.

Orestes H. Saint John, another of Agassiz's assistants on the Brazilian journey, is the only student of fossil fishes that Agassiz had in America. An extremely modest man, Saint John has since distinguished himself by the publication of memoirs of great value on the fossil fishes of Illinois and other Western states, and stratigraphical researches on the state of Kansas.

As to William James, a gifted member of a gifted family, his journey to Brazil and up the Amazons developed his keen power of observation, in a psychological direction and in the philosophical realm, and he has since been made one of the professors of philosophy at Harvard University, and become one of the most enlightened disciples of the Charcot school of psychology.

Samuel H. Scudder, the author of the charming sketch, "In the Laboratory with Agassiz," is one of the best pupils, and perhaps the most devoted naturalist, who studied under Agassiz. He has devoted his life to American entomology, living and fossil, and has published standard and most beautiful works on these subjects. Fossil insects, more especially, have been his favourite study for many years, and we may truly say that he is the American palæo-entomologist *par excellence.*

John B. Perry came to Cambridge among the last of Agassiz's pupils and assistants. I had met him at Swanton, Vermont, in 1861, where he was a pastor of the Congregational church, and was much impressed by his capacity as an observer in practical geology. He was my constant companion during all my researches on the Taconic system, in Vermont and Northeastern New York, and, as his biographer says, my "friendship was the great turning-point in Mr. Perry's future." He rapidly became a good palæontologist, and in 1868 left the ministry to accept a position as assistant in the Agassiz Museum. But Perry did not live long; during a protracted excursion in the Southern States, during the summer of 1872, in search of Tertiary fossils, he contracted malaria, and died of it at Cambridge in October, 1872. He was a man whose honest and modest diligence as a geologist Agassiz highly appreciated.

During the fall of 1867 Agassiz lost his mother, — the heaviest sorrow of his life. She died at Montagny on November 11, at the age of eighty-four years. "Madame la pasteur Agassiz," as she was called in Switzerland, was a most remarkable lady, superior to all her surroundings. Every one loved her, and she was respected as few women ever were. Her son Louis was her favourite child, and in her Agassiz found a profound maternal love, comforting him in all his trouble, giving gentle counsel, never discouraged, but always hoping for better times to come. Mother-like,

she kept her interest in all his work in America, looked
over all his publications in the English language, although she did not understand English, and every volume and every paper received from the other side
of the Atlantic from her dear Louis was carefully
treasured by her.

CHAPTER XXI.

1868–1870.

The activity of Agassiz in scientific lines knew no bounds, as will be seen from the following extracts from some of his letters to me, written during 1868.

CAMBRIDGE, MASS., le 13 mars, 1868.

Mon cher Marcou, — Vous savez peut-être que Mr. Peabody, le banquier américain à Londres a donné à notre Université une somme de 150,000 dollars pour fonder un musée ethnographique. Tôt après cette donation je me suis mis en campagne pour engager les "Trustees" au nombre desquels est Mr. R. C. Winthrop, aujourd'hui en Europe, à faire quelque grande acquisition comme point de départ. J'ai d'abord recommandé la collection du Colonel Schwab (Switzerland) ; je vais presser celle (de Gabriel de Mortillet) que vous, me recommandez, et si le Musée était déjà organisé je ne doute pas que la chose ne se fît, mais tout est encore en question, jusqu'à l'emplacement à choisir pour y déposer les collections.

163

Votre lettre du 24 février a trait à mes "Principles of Zoölogy."
Je n'ai jamais reçu de communication à ce sujet, ni de votre ami
M. Reclus, ni de son éditeur M. Hetzel. Ce sont de ces choses
auxquelles on n'est pas indifférent et je verrais avec plaisir ce livre
traduit en français; mais il a besoin d'être revu. Plusieurs cha-
pitres sont vieillis et bien qu'il en ait paru plusieurs éditions ici, je
n'y ai jamais fait de retouches.

Votre dévoué,

Ls. AGASSIZ.

The Hon. Robert C. Winthrop visited me in Paris,
and asked my help in the purchase of de Mortillet's collection. The transaction was successfully made, and I
superintended the packing of the collection, after receiving it into my charge, happy to have another opportunity to help Harvard University. The elementary
treatise on zoölogy, although translated by Elisée
Reclus as far back as 1867, was not published until
1891, when it appeared in Hetzel's "Magasin d'Education et de Récréation," under the title, "Principes de
Zoölogie, par L. Agassiz et Gould, traduit par Elisée
Reclus." This reprinting of the work forty-three years
after its original publication shows how highly it was
appreciated, and is unique in the case of an elementary
book, treating of so variable a science as zoölogy.

NAHANT, le 4 juillet, 1868.

Mon cher Marcou, — Vous aurez peut-être appris que j'ai fait
une grosse maladie qui a failli m'enlever et dont je me remets
lentement. Toute application m'est interdite; seulement je voudrois, s'il en est encore temps, m'assurer la collection de Poirrier.
Veuillez dire à sa veuve que si elle n'a pas conclu avec Mr. Cope, je
la prie de me donner la préférence, puisque j'ai été en négociation
pour cette collections avant lui.

Encore une demande. La Législature [of Massachusetts] vient

de m'accorder une forte somme pour étendre le Musée et je voudrois y attacher un mouleur de première force. Voudriez-vous voir Stahl au Jardin des Plantes et lui demander s'il serait disposé à venir me rejoindre ; et si non, peut-être a t'il formé quelqu'élève qu'il pourrait me recommander consciencieusement pour faire des travaux délicats, aussi bien que des moules de grands ossements fossiles. Faites lui en même temps mes amitiés.

Dès que je serai moins affecté par la position nécessaire pour écrire qui me suffoque, je reprendrai la plume plus longuement.

Bien à vous,

Ls. AGASSIZ.

This letter shows unmistakable signs of the difficulty Agassiz experienced in holding his pen. Indeed, I never received one from him so plainly indicating tremulousness.

His health, regained during his Brazilian journey, broke down again during the spring of 1868. The heart was affected this time, and Dr. Brown-Séquard ordered a complete cessation of cigar-smoking, — a great privation to Agassiz. He was not an easy invalid, being too full of schemes of all sorts, and very impatient of bodily inactivity.

The Poirrier collection, referred to in the above letter, was a rich collection of very rare vertebrates from the fresh-water Tertiary formation of Auvergne. I had to look at it attentively, at the request of Agassiz ; but the death of M. Poirrier and the serious sickness of Agassiz prevented the purchase. The collection has been secured since by Mr. Cope, of Philadelphia, and, consequently, is in good hands.

M. Stahl, Agassiz's former modeller at Neuchâtel, trained with great care to the work of modelling fossil

fishes and other objects, had become so expert that the
Jardin des Plantes of Paris would not part with him;
and besides, he felt, as he told me, that he was too old
to begin life again in America. However, Agassiz,
with his usual persistency when he wanted anything
connected with natural history, got a modeller, and created at his museum a studio where numerous specimens, especially of fossil bones, were modelled. But it
proved too expensive, and at his death the studio work
was discontinued.

UTICA, N.Y., le 10 août, 1868.

Mon cher Marcou, — Vous savez peut-être que j'ai fait une maladie grave et douloureuse l'hiver dernier. Ne me remettant pas
assez vite pour pouvoir espérer de reprendre mon travail l'hiver
prochain en restant à Nahant, où je suis harcelé de visites, je me
suis mis en route pour les Montagnes Rocheuses, et me voici à
Utica, où je passe le dimanche, occupé à étudier votre Géologie et
carte géologique de l'Amérique du Nord.

Pendant mon voyage je m'appliquerai surtout à étudier l'influence
que les Montagnes Rocheuses ont eue sur les phénomènes erratiques
durant l'époque glaciale; c'est un point de la question qui n'a pas
encore été touché. J'étudierai aussi le lit de nos fleuves, dont les
dimensions se rattachent aussi à cette question. Comme je l'ai été
à l'origine, je veux, si possible rester à l'avant-garde pour tout ce qui
touche à ce sujet. Quoiqu'on cherche à la rapetisser, c'est après
tout une des plus grandes questions de la géologie et quoiqu'en dise
les envieux c'est moi qui l'ai soulevée et développée.

La part de Charpentier est d'avoir mis en évidence la plus grande
extension des glaciers des Alpes. Il ne s'est jamais avisé d'y voir
un phénomène général indiquant des changements profonds dans la
température du globe. J'ai été surpris de voir que bien des géo-
logues sont disposés à lui faire honneur aussi de ma part de la
question. . . .

Adieu, mon cher **Marcou,** au revoir.

LS. AGASSIZ.

This letter is doubly interesting: first, because it shows that the great and wonderful memory of Agassiz was on the decline, for he had forgotten his preceding letter in regard to the state of his health ; second, because it expresses his feeling about the glacial question and the studied neglect of some, who affected to pass over the great part he had taken in it, from the start.

In 1868, a short journey.to the Rocky Mountains was made on the invitation of Mr. Samuel Hooper, then member of Congress for Massachusetts, to see the progress made in the construction of the Union Pacific Railroad.   General Sherman, commander-in-chief of the army, joined the party with ambulances and an escort of cavalry for conveyance across a part of the Kansas and Nebraska prairies.   The road was built only as far as the Green River station, in Wyoming, where the laying of rails had brought to view a bed of limestone full of fossil fishes and insects.   Agassiz was delighted by the discovery, and, if he had been fifteen years younger, he would never have stopped at Green River station and thence turned eastward without a visit, strongly urged upon him by Judge Carter, of Fort Bridger, to the Grizzly Bear Buttes, so celebrated since as the richest locality for fossil vertebrates of the Oligocene period.   Judge Carter, the sutler at Fort Bridger, met the party at Green River, and related to Agassiz, how an old trapper, Jack Robinson, had repeatedly reported to him that he knew several places, at the foot of the Uintah Mountains, where grizzly bears were changed into stone, the bones being as hard and heavy as rocks ; and, in order to convince Judge Carter, the trapper one day brought a bag full of these fossil bones, which he threw down on the floor at his feet. To his surprise, he saw a well-preserved skull, resembling that of a bear, which has since been described by Dr. Leidy as *Palæosyops.*

If Agassiz had accepted the hospitality and help most generously offered by Judge Carter, he would have anticipated the disoovery, made the year after, in 1869, by Hayden's exploring party. But Agassiz was too old and not strong enough to undertake a horseback ride of sixty miles to reach Fort Bridger. His experience on horseback the year previous, in the province of Ceará, in Brazil, had entirely disgusted him with this mode of conveyance, and he firmly declined Judge Carter's invitation.

This is another instance of the singular case of atavism of his ancestors, the lake-dwellers of Switzerland. Like them, Agassiz would go on any kind of boat, however unsafe it might be; but on land he trusted only his legs.

When, in 1875, during an exploration round Fort Bridger, I told Judge Carter of the strong aversion Agassiz had to riding horseback, he exclaimed, "Oh! if I had been aware of it, I would have brought with me an ambulance, for I wanted so much to show him the specimens."

On his return to Cambridge, Agassiz stopped at Cornell University, which was about to be inaugurated, and where he had accepted an appointment as non-resident professor. He assisted at the opening of the University, and made one of his characteristic speeches, entering most heartily into the new enterprise, pledging it his support, and giving most valuable hints as to the proper lines of development. With his optimistic tendency, Agassiz took a great fancy to the idea of a university combining the artisan with the student, manual labour with scientific work. He came back from Ithaca, the seat of the Cornell University, with the most exaggerated views in regard to the future of the new institution, speaking of the backwardness of Harvard, and prophesying that Cornell would soon leave Harvard far behind. In his enthusiasm for a new plan, Agassiz was apt to go too far. He seemed to forget entirely that an old university like Harvard must always possess an amount and kind of interest which a new university cannot have. A university is the work of time; and money and new plans cannot take the place of the long series of accessions, material as well as intellectual, which has enriched Cambridge during more than two centuries.

A letter to Agassiz from the poet Longfellow, then travelling with his family in Europe, gives me an opportunity to recall several facts, all to the honour of Agassiz, as contributions, coming either from scientific adversaries like Darwin and Huxley or from admirers like Tyndall. On meeting Longfellow at the Isle of Wight, Darwin said to him, "What a set of men you have in Cambridge! Both our universities [meaning Cambridge and Oxford] put together cannot furnish the like. Why, there is Agassiz—he counts for three." Coming from Darwin, the compliment is no small one, for Agassiz had opposed his "Origin of Species" in undisguised terms for eight years. A few months after, at the Athenæum Club in London, Thomas Huxley said to me, "Agassiz is a backwoodsman in natural history. He clears up the forest, cutting down all errors, theories, without regard to persons or established reputation. What a pioneer!"

Some years before, at the meeting of the Swiss naturalists at Geneva, in 1865, I heard John Tyndall say, "If Bishop Rendu had been a physicist, he would have left nothing for me to do; for after the experiments of Agassiz on the Aar glacier, all the main facts of the glacier motion and mechanism were so well established as to leave nothing but a question of pure physics, which was as nearly as possible solved by Rendu's theory." Tyndall expressed, without reserve, his admiration of Agassiz's work, and his dissent from James Forbes's theory and claims in regard to the structure of ice.

Agassiz was never a good judge of character; and he too often associated with himself persons either unfit for the work assigned to them, or not in a condition to render the services expected. He was too easily led by flattery, and was apt to trust any one who made a show of devotion to the progress of science, and spoke grandiloquently of the sanctity of the sacred office of carrying on researches in various scientific fields. He was imposed upon by the airs assumed by a certain number of half savants, who are to be found everywhere, but in greater number in America, where they have received the name "Almighty Savants."

His professorship of zoölogy and geology, on his

proposition, was successively divided into three separate chairs; a chair of geology, a chair of palæontology, and a chair of zoölogy.   At the time of his death neither of the incumbents were persons well fitted for the position for which they had been chosen.

The first mistake was the calling to the chair of geology of an indifferent observer.   We know how Agassiz was justly proud of his knowledge on glaciers and of being considered as the father of the "Ice-age." What did his successor do but publish, in 1882, a long and diffuse paper entitled, "The climatic changes of later geological times," in Volume VII. of the "Memoirs of the Museum of Comparative Zoölogy," founded by Agassiz, in which he attempts to nullify his greatest discovery in geology.   It is almost incredible that at this time of our knowledge of glaciers and the glacial question a person has called the "Ice-age" a myth, saying, pp. 387 and 388 in the paper quoted above: "The so-called glacial epoch . . . the glacial epoch was a local phenomenon," just the reverse of the discovery and teaching of Agassiz.   The same person, after ascending Mount Shasta in California, and exploring the Sierra Nevada, has the boldness to emphatically declare that no glaciers exist now at either Mount Shasta or in the Sierra Nevada, where we all know they may be counted by the half-dozen.

The second mistake was his choice for the chair of palæontology.   Agassiz formed some pupils who greatly honour his teaching in palæontology.   It is sufficient to name such able American palæontologists as Alpheus Hyatt, Samuel H. Scudder, and Orestes St. John.   But instead of choosing one of these, he appointed another pupil, who after a futile attempt at work on the Brachiopoda, took to teaching geology. After some years his official position of professor of palæontology became so embarrassing, even to himself, that he asked to have his title changed to professor of geology.

Finally, the third mistake was the appointment of a third-rate zoölogist for the chair of zoölogy. After a few years the incumbent retired.

The preceding details are necessary in order to show a foible in Agassiz's character, and how some of his greatest efforts and successes were partly paralyzed by his choice of associates and substitutes in zoölogy, palæontology, and geology.

It had become an absolute necessity to increase the financial resources of the Museum; for the constant addition of specimens involved such expense that it was almost impossible to carry on active operations even in the most meagre way. In 1867, Agassiz obtained a grant of ten thousand dollars from the Legislature of Massachusetts, and from the American Congress the remittance of excise duty on alcohol used for scientific purposes; and again in 1868, the Legislature granted seventy-five thousand dollars for an extension of the building, and private individuals subscribed a similar sum. Work was begun at once, and two-fifths of the north wing was added to the two-fifths already standing. It was impossible to do more at the moment, and the final fifth of the north wing was left to be added at some future time.

Over four hundred and seventy-three thousand dol-

lars had been expended on the Agassiz Museum from
the time of its organization in 1859; certainly a large
sum, but not too large, if we consider the results arrived
at. As Agassiz justly says: "It is an astonishment
and a gratification to me to find that in ten years we
have attained a position which brings us into the most
intimate relations with the first museums of Europe;
we have a system of exchanges with like establishments over the whole world; while the activity of
original researches in our institution, and its well-sus-
tained publications, the possibility of which we owe to
the liberality of the Legislature, make it one of the
acknowledged centres of the scientific progress. . . .
I claim that its results, as compared with those of
other institutions, are in more than due proportion to
the money expended. . . . The organization must, of
course, be the work of the director; but for the energetic and intelligent carrying out of the scheme, I
have to thank the gentlemen working with me either
as assistants upon very moderate salaries, or as friends
of the institution who give their work without any
renumeration whatever.[1] . . . From the earliest organization of the Museum, I have had three great objects
in view. First, to express in material forms the present state of our knowledge of the animal kingdom;
second, to make it a centre of original research, where
men who were engaged in studying the problems connected with natural history should find all they needed
for comparative investigations; thirdly, — and this last

---

[1] "Report of the Director of the Museum of Comparative Zoölogy" for
the year 1869.

object has been by no means less prominent than the
two others, but, if possible, has engrossed my thoughts
more, — to make it an educational institution; to give
it a widespread influence upon the study, the love, and
the knowledge of nature throughout the country. . . .
I have laboured under many obstacles in the carrying
out of this scheme. Often, for want of means to pay
salaries, the assistants have been so few, and their
knowledge so immature, that it was impossible to
organize any extensive scheme of instruction." [1]

During the spring of 1869 Agassiz joined Pourtalès
on the United States coast survey steamer *Bibb*, engaged in deep-sea dredging between Florida, Cuba, and
the Bahama Islands.   Dr. William Stimpson, a favourite
pupil of Agassiz, had inaugurated dredging for marine
animals along the New England coasts, but to Pourtalès
are due the systematic investigations of the beds of the
Atlantic Ocean, on the American side, having for their
aim the fauna existing at different depths.   In the two
years 1867 and 1868, Pourtalès had succeeded so far as
to leave no doubt that "animal life exists at great depths
in as great a diversity and as great an abundance as in
shallow water."

Agassiz in his " Report upon Deep-sea Dredgings in
the Gulf Stream, during the Third Cruise of the United
States Steamer *Bibb*," Cambridge, November, 1869, says,
pp. 363 and 367 : " The object of my own connection with
the present cruise was to ascertain how far the last investigations covered the ground to be surveyed, and to what

[1] "Report of the Director of the Museum of Comparative Zoölogy " for
the year 1868.

extent and in what direction further researches of the kind were desirable in the same region, and likely to furnish important information. . . . It is a pleasure to me to state that our cruise — extending farther to the east in the Gulf Stream, between Cuba and the Bahamas on one side, and Florida on the other, than those of previous years — confirmed in every feature the conclusion already reached by M. Pourtalès. . . . Permit me a suggestion. . . . It would be appropriate and just that this extensive coral plateau, the characteristic fauna of which M. Pourtalès has so faithfully explored, should bear his name and be called the 'Pourtalès Plateau.'"

The Boston Society of Natural History proposed to celebrate the centennial anniversary of the birth of Alexander von Humboldt on the 14th of September, 1869, and appointed Agassiz orator of the day. He accepted the invitation with joy, and was grateful for the honour, because, as he says, he "loved and honoured the man." The address was delivered at the appointed time, in the Music Hall of Boston, before a crowded and brilliant audience comprising many persons of the highest culture and distinction in New England.

The choice of Agassiz was eminently proper, not only on account of the nature of the work done by Humboldt in the New World, but also because of Agassiz's personal intercourse with him, which began when Professor Oken wrote from Munich in 1829 to offer Agassiz's services as an assistant for Humboldt's journey in Central Asia. It was more especially during Agassiz's

stay in Paris, 1832, that he saw much of von Humboldt.
Then the great traveller was at the zenith of his reputation and social success. He was the lion of all the political, literary, and scientific salons of the French capital.
Humboldt was an admirable talker. He would con-
verse for hour after hour, hardly taking time to breathe,
keeping the whole circle of hearers collected round him,
and hanging in suspense on his lips. He was a monol-
ogist *par excellence ;* few people were bold enough to
interrupt him, even by an exclamation of admiration
or wonder. An anecdote will best show the great
attraction exercised by von Humboldt upon all savants.
His friend and companion at the School of Mines of
Freiberg, Leopold von Buch, was so fond of hearing
him that he could find nothing better to do than to
waylay him as he left " les salons," during the winter
of 1820 in Paris, walk home with him, and sit down and
listen to him all night, from midnight until daybreak.
The result was that von Buch suffered from an attack
of pneumonia, and when reproved by his cousin, the
young Count d'Arnim, for his imprudence, the answer
was, "It is my fault. The open fire near which we
were talking went out. I was very cold and chilly, but
if I had made the slightest move to light it again, perhaps I would have caused Humboldt to leave me. I
preferred to suffer and hear his conversation, and I am
very glad of it, because I have gained much knowledge
by it."

The only person in Paris who treated Humboldt as
an equal and did not fear to interrupt him and even to
make fun of him, was the great astronomer, François

Arago. One.day he said to him before witnesses, one of whom repeated it to me, " You do not know how to write a book. When you begin a subject, you go, go, go, and cannot stop, just like your never-ending talk." ["Toi, Humboldt! tu ne sais pas écrire un livre; tu commences, va, va, va, tu ne peux plus t'arrêter; exactement comme tu parles."] Arago always *tutoyait* Humboldt, a custom dating from the French republic at the time of the *sans-culotte*, which was kept up for one or two generations among the savants, and even to this day among artists.

The address of Agassiz was admirable. It gave all the salient points of Humboldt's life; indicated his great influence on the progress of natural sciences during the first half of this century, and acknowledged the debt America owes to him as a discoverer in physical geography, — a science which it may be said was created by him, — and to his clear and exact exposition of everything relating to the natural history of the equinoctial regions of the New World. Humboldt was inclined to be sarcastic and was always ready to make fun of others, the only ones exempt from his rather sharp remarks, among his scientific contemporaries, being Arago and young Agassiz. Even the glacial doctrine, which he did not relish much, was treated respectfully by him, and he used it only against his old friend von Buch, who always lost his temper every time a reference was made to it.

To deliver his address, Agassiz had to leave the side of his dear and favourite child, Mrs. Pauline Shaw, then very ill and in a critical condition. The great effort of preparing his elaborate address, its delivery before a large audience, combined with the sorrow due to his daughter's illness, proved too much for his brain; and soon after he broke down completely, and a very severe and dangerous attack of paralytic apoplexy disabled him for more than ten months. His speech was affected, he was unable to control his hands and hold anything, and physicians forbade every exertion, even thinking. This last privation was the most painful, and not easy to bear without constant desire to break the doctor's order. Lying on a lounge in his sick-room, he looked like a lion loaded with chains and encaged in an iron box. His splendid and strongly built body was no longer at the command of his will; his inquisitive, brilliant, and intelligent eyes followed closely every visitor, as if to inquire what they thought of his sickness. Vertigo was a constant menace. However, as soon as he could, he began to dictate notes in regard to the arrangement of the more recent collections received in his museum; and he called to his side some of his assistants, to confer with them regarding lectures and the application of a new and strict rule to each employee, compelling each to work seven hours a day purely for the benefit of the institution, no outside work, even of a scientific character, being allowed during that time.

In the spring of 1870, as soon as it was possible to remove him without too much risk, he left Cambridge for the small village of Deerfield, on the Connecticut River. There he improved rapidly; the vertigo symptoms soon disappeared, daily walking about the village

being the only "cure" he followed, with the relish of
an old walker, accustomed since his life at Motiers,
Orbe, and Concise to ramble in search of animals
and plants.   As soon as he could write, he began a
correspondence with the trustees and his assistants at
the Museum and his old scientific friends in Europe,
showing how miraculously his health had recuperated,
and activity returned into his frame, like the rebound-
ing of an elastic ball.   In November, 1870, he returned
to his dear museum, and was able to resume his lectures.

Although Agassiz never meddled, or even troubled
himself much, with politics, he followed with an intense
interest the Franco-German War.   The abuse of victory
by the new German emperor particularly wounded him ;
he thought better of a Prussian king, and from the
moment he read the autocratic and exacting terms of
the peace forced on France by an ungenerous and
impolitic victor, he turned against Germany, all his
old French instincts and early impressions of the great
services rendered to liberal ideas and science by France
and Frenchmen coming back to his memory, and blot-
ting out all the sympathies aroused during his student
life in German universities.   It was a complete rever-
sion of sentiment, and he at once wrote a letter full of
sympathy to his old friend, M. Thiers, then president
of the French republic, receiving a most flattering letter
of thanks some months later.   Agassiz knew too well
what was due to French influence in every department
of human knowledge to accept all that was said against
France and the French nation.   His original Vaudois
blood revolted, and from that moment, until his death,

he spoke openly of his sympathies and ardent feeling for a people which had rescued his native Canton de Vaud from Bernese tyranny and bondage, and is of the same Latin race, using the same language, — the language of Jean Jacques Rousseau, de La Harpe, Jomini, and Madame de Staël, and in which his own works on the "Poissons fossiles" and on the glaciers were written. The crisis brought up by the fall of France, added to old age, made him, by a sort of irresistible tendency, return naturally to the feelings of youth and of childhood, while those of middle age disappeared, and after 1871 Agassiz was more French than myself, at least in his feeling about the future of France.

CHAPTER XXII.

1871–1872.

PUBLIC life possesses such an attraction, and gets
such a strong hold on any one who has drunk at the
cup of popularity, that it is almost impossible to resist
the temptation of maintaining a name which has once
become celebrated on the world's stage. Great naturalists, like statesmen, artists, speculators, financiers, are
no exception to the rule. Simple prudence, after the
illness of 1870, would have easily prolonged Agassiz's
life for a score of years, for he came from long-lived
parents, his mother dying in the eighty-fifth year of her
age. But Agassiz was not a man to step into comparative obscurity; he wanted applause, not only in the
lecture-room, but also before the general public. The
discovery of animals living at great depth on the bottom of the sea was so interesting that he was unable to

resist the desire for an investigation directed by himself in person.

His friend, Professor Benjamin Pierce, had succeeded Professor Bache as Director of the United States Coast Survey, and an expedition was easily arranged under his supervision. Frank de Pourtalès, who had passed the last five years in deep-sea dredgings for the Survey, was naturally put in charge of the apparatus for sounding and hauling the net from the bottom of the sea. A new steamer, built especially under the direction of the navy officer, Captain C. P. Patterson, chief hydrographer of the Coast Survey, and named the *Hassler* in honour of a Swiss mathematician of Aarau and the first director of the Coast Survey, was fitted for the voyage. It was a small steamer of three hundred and fifty tons, rigged as a three-masted schooner, one hundred and sixty-five feet long, twenty-five feet beam, ten feet depth of hold, and with a draught seven and a half feet forward and ten feet aft; but it was too hastily and imperfectly finished by the contractor, and was not a proper vessel for a long voyage.   The compound engines, with double cylinders, the same used at that time on the White Star Line Company's transatlantic steamers, were unfit, and ought never to have been accepted.   A lack of oversight, during the construction and at the reception of the little steamer, marred the whole expedition from its start until its arrival at San Francisco.   It was a great, almost a cruel, carelessness to embark a man so distinguished, so old, and so much an invalid as Agassiz was, in an unseaworthy craft, sailing under the United States flag.

The crew and passengers numbered about fifty persons, the scientific party being composed of Professor Agassiz and his wife; Dr. Thomas Hill, ex-president of Harvard University; Frank de Pourtalès; Dr. Franz Steindachner, Agassiz's favourite assistant in ichthyology, and a draughtsman; and Commander P. C. Johnson, United States Navy, who was also accompanied by his wife, a native of Chili. Everything that could contribute to the comfort of Agassiz and his friends was provided.

After long delays, from the summer until the beginning of the winter, the *Hassler* at last sailed from Boston, the 4th of December, 1871, in a snowstorm, and with a heavy sea. During the first four days, the roughness of the voyage much fatigued Agassiz, who thought that he had undertaken a task beyond his physical strength. However, after passing Cape Hatteras and reaching the West Indies, he quickly rallied from the discouragement and depression which had seized him at the departure; and a very successful dredging, made in a depth of eighty fathoms near the coast of the Barbadoes, which brought up several stemmed crinoids, then a great rarity, delighted him so much that he soon forgot all the unpleasantness of the voyage.

Unhappily, the station near the island of Barbadoes was the only one of the whole expedition which was successful so far as number of sea-animals obtained was concerned   It soon became evident that the engine of the *Hassler* was not only defective, but absolutely worthless; and the whole plan of the voyage, sketched with such a masterly hand, had to be modified and curtailed. The *Hassler* was able merely to creep along the coast from one port to another, stopping at almost every one to allow her machinery to be repaired. She sailed from the Barbadoes to Pernambuco, thence to Rio Janeiro, Montevideo, Port of San Antonio, Strait of Magellan, Arenas, Port Famine, Glacier Bay, Sholl Bay, San Pedro, and finally to Talcahuana in Conception Bay, Chili, where she remained three weeks for the repair of her engine. Pourtalès dredged as often as practicable, and succeeded in collecting a large number of rare or unknown species.

Before embarking, December 2, Agassiz wrote " A Letter concerning Deep-Sea Dredgings, addressed to Professor Benjamin Pierce, Superintendent United States Coast Survey,"[1] in which he says: —

On the point of starting for the Deep-Sea Dredging Expedition, for which you have so fully provided, and which I trust may prove to be one of the best rewards for your devotion to the interests of the Coast Survey, I am desirous to leave in your hands a document which may be very compromising for me, but which I nevertheless am determined to write, in the hope of showing within what limits natural history has advanced toward that point of maturity when science may anticipate the discovery of facts.

If there is, as I believe to be the case, a plan according to which the affinities among animals and the order of their succession in time were determined from the beginning, and if that plan is reflected in the mode of growth, and in the geographical distribution of all living beings ; or, in other words, if this world of ours is the work of intelligence, and not merely the product of force and matter, the human mind, as a part of the whole, should so chime with it that, from what is known, it may reach the unknown ; and if this be so, the amount of information thus far gathered should, within

---

[1] "Bulletin Mus. Comp. Zoölogy," Vol. III., Cambridge, 1871.

the limits of errors which the imperfection of our knowledge renders unavoidable, be sufficient to foretell what we are likely to find in the deepest abysses of the sea, from which thus far nothing has been secured.

. . . There is a correlation between the gradation of animals in the complication of their structure, their order of succession in geological times, their mode of development from the egg, and their geographical distribution upon the surface of the globe. If that be so, and if the animal world designed from the beginning has been the motive for the physical changes which our globe has undergone, and if, as I also believe to be the case, these changes have not been the cause of the diversity now observed among organized beings, then we may expect, from the greater depth of the ocean, representatives resembling those types of animals which were prominent in earlier geological periods, or bear a closer resemblance to younger stages of the higher members of the same types, or to the lower forms which take their place nowadays.

As the adversaries of Agassiz not only sharply criticised, but even scouted these views, and cast all manner of fun upon them, and as I made the suggestion to him that successive marine faunas would be found at successive great depths, according to the number of fathoms, showing a correlation between the depth and the geological periods, the deepest possessing forms of the primordial fauna, and consider myself as responsible for it, I would call attention to the following facts, very lately made public.

This prophetic announcement has been at least fully confirmed in regard to *Radiolaria*, which have been brought up from great depths, varying from six thousand to thirteen thousand metres, and present numerous genera absolutely identical with forms existing, not only in the Mesozoic and Palæozoic strata, but even from the Infra-primordial fauna of the Lower Taconic. The persistence of several genera from the oldest quartz-ite rocks in North Britanny, near Saint Lô,[1] through millions of years, notwithstanding the metamorphoses of all other animals around them, is a fact which cannot be put aside by transformists and Darwinians. Immu-tability of several genera, from the beginning of life on our planet until now, is not in favour of natural selec-tion or evolution. Permanence of forms has existed since the first appearance of life on our planet, — a privilege only enjoyed now by *Radiolaria* and perhaps by *Foraminifera*, but which may be extended, as Agassiz thought, to higher animals, such as Trilobites and Am-monites. It seems only a question of time; for we know yet so little of life at great depths in our oceans that some unexpected discovery may be made, and prove a great bar to all the hypotheses constantly resorted to by the theorists of the Darwin school.

If the hopes formed by Agassiz were not fulfilled by the *Hassler* expedition, it was due mainly, first, to the defects in the dredging apparatus; second, to the inadequate estimate of the time required to explore the deepest abysses of the sea. We shall not know for centuries to come all the fauna of the deep sea. The difficulties of finding animals and bringing them from the sea-bottoms, at great depths, are such that centu-ries will be required before a complete knowledge of

---

[1] " Les preuves de l'éxistence d'organismes dans le terrain précambrien. Première note sur les Radiolaires précambriens," par L. Cayeux (" Bulle-tin Soc. Géol. France," 3d Série, Tome XXII., pp. 197–228; Paris, 1894).

the exact distribution of marine animals at the different zones of depth in the oceans can be expected.

The most interesting geological parts of the voyage, especially for Agassiz, were the visits to glaciers in the Strait of Magellan and in Smithe's Channel. There he found, almost at sea-level, great glaciers like the Aar, Aletch, and Rhone glaciers of Switzerland. The marks of *Roches moutonnées*, moraines, scratched pebbles, and boulders were seen, and ancient traces of glacial action found in many places, on the southern extremity of the South American continent, which was a great satisfaction to the old landlord of the " Hôtel des Neuchâtelois " in the Bernese Oberland; and the discoverer of the "Ice epoch" had the pleasure of seeing his prediction in his Discourse at Neuchâtel in 1837 verified, even in the Southern Hemisphere. As he said on his return to Cambridge, " If I had done nothing else but see and study the glacial phenomena at the Magellan Strait and among the Chiloë Islands, this would have amply paid me for all my trouble and fatigue." What does not geology owe to such an observer? To him, and to him alone, is due the discovery of the existence of glaciers in Scotland, England, and Ireland, and their extension over all New England, and in the province of Ceará in Brazil, and in Chili. Seldom does a savant have the opportunity to verify, on the field, in both hemispheres, observations made in the limited area of such a small country as Switzerland. Such a success was only possible to the foresight and genius of an Agassiz.

At Talcahuana Agassiz disembarked, and thence

travelled by post to Curicu and by rail to Santiago. The *Hassler*, after repairs, resumed its cruise to the island of Juan Fernandez, with only Pourtalès and Dr. Hill on board, to continue the dredging and other scientific observations. But the sounding-lines broke down entirely, the ropes rotted, and it became impossible to dredge at a depth even of one hundred fathoms.

It was a great disappointment to Agassiz and Pourtalès, after coming so far, to be deprived of expected results by defective machinery and worthless apparatus. Before leaving Cambridge, Agassiz had written in his Annual Report as director of the Museum for the year 1871 that he was going "to explore the greatest depths of the Atlantic and Pacific oceans, on both sides of the American continent"; and in no case was a really great depth reached by the dredge.

At Santiago Agassiz met two old European friends, both eminent naturalists, — Don Ignacio Domeyko, rector (president) de la Universidad de Chile, and Dr. R. A. Philippi, professor of zoölogy and botany at the same university. Domeyko was a student at the Paris School of Mines, when Agassiz was there in 1832; and both met at Cuvier's house and in the office of Élie de Beaumont. No one has done so much as Domeyko to develop mining in Chili, and no one was held in such high regard and respect all over the Chilian Republic. He received Agassiz with open arms, and he and his daughter Annita did all they could to make his visit agreeable and profitable. Philippi had visited Agassiz at Neuchâtel on his return from a scientific exploration of Sicily with Friedrich Hoffmann and Arnold Escher von der Linth, and there had been a friendly intercourse between them ever since that time. Philippi went to South America about the time that Agassiz came to North America, and his exploration of the great desert of Atacama and subsequent publications are justly celebrated. The alcalde and municipal body of Santiago called on Agassiz, tendering an invitation to a great dinner party and reception, in his honour; but his precarious health and fatigue from the journey prevented him from accepting.

A very agreeable telegram from the Emperor of Brazil awaited Agassiz at Santiago; it announced his election as a "Membre étranger de l'Académie des Sciences de l'Institut de France," an honour seldom conferred, on account of the limited number (eight) allowed in this class of members. Several times before, his name had been on the list presented by the committee for election, but, curiously enough, he was always opposed by the zoölogists, while his just claim to the distinction was strongly supported by the physicists, astronomers, mathematicians, botanists, and medical fellows of the Academy. As he says, in a letter to Dom Pedro Secundo : "The distinction . . . unhappily, is usually a brevet of infirmity, or at least of old age, and in my case it is to a falling house that the diploma is addressed. I regret it the more because I have never felt more disposed for work, and yet never so fatigued by it."

He joined the *Hassler* at Valparaiso, after a few days of rest at Santiago, and continued his voyage.

Some dredgings in shallow waters (for the lines had
been too short to allow any other dredging) were successfully made along the coast of Peru, and then the
steamer headed for the Galapagos Islands.

Ever since Darwin's exploration of these islands in
1835, Agassiz had had a great desire to see the home
of the lizard *Amblyrhinchus*, that remarkable remnant
of Secondary or Mesozoic times, and he now had the
pleasure of watching its capture in considerable numbers, as Steindachner and Pourtalès hunted it on the
rocks and in the shallow waters; while from the deck
he drew their attention to a large specimen lying on
the sand, half choked by the pocket handkerchief tied
round its neck, and which swiftly turned round, as it
revived, ready to plunge into the sea.   Specimen after
specimen was placed in alcohol; for Agassiz, as usual
with him in the case of rare animals, was never satisfied.   The collections made in the Galapagos were
important and very valuable, but Agassiz was too old
to obtain the full benefit of such an exploration; if it
had been made in the prime of his scientific life, the
result would have been different.   Sickness tormented
him during his stay.   However, he was quite well again
when the steamer dropped anchor opposite Taboga, in
the bay of Panama.   As soon as his arrival was known
in town, fishermen and pearl traders came on board, the
latter to find with surprise that a naturalist of such reputation was not necessarily a great purchaser of pearls.
His hands were filled with the most superb specimens
collected in the bay, several being very large and perfect;
but, to their astonishment and disgust, Agassiz pre-

ferred to purchase the most common fishes brought from the city market.

Here the party separated. Pourtalès crossed the Isthmus and took ship at Colon for Washington, while Agassiz and the other members of the party cruised along the Pacific coast of Central America, Mexico, and the two Californias, entering the Golden Gate of the bay of San Francisco on the 24th of August, 1872, a little less than nine months after leaving the wharf of the Charlestown navy-yard at Boston.

A month was spent at San Francisco, in social activity and enjoyment of the great metropolis of the Pacific states, although Agassiz, who was much in need of repose, declined all invitations of too exciting a nature. Every one showed much kindness to both Agassiz and his wife; in fact, it was impossible for Mrs. Agassiz even to go shopping without being recognized by tradesmen and their clerks. Agassiz was too weary of travelling to undertake any scientific researches round San Francisco, and he did not even visit the gigantic Sequoia trees in Calaveras County.

Early in October Agassiz found himself once more in his museum at Cambridge, and it was a great pleasure to the assistants and to all the friends of the institution to see his genial face again. It seems in place here to complete what has been previously said in regard to his museum.

The first four years of its existence, from 1860 to 1864, were very difficult years. Its success was all that could be reasonably expected, both as to its exhibits, which were tolerably presentable, and as to its scientific arrangement. And as it was, the governors of Massachusetts, — Banks and Andrew, — the members of the Legislature, numbers of savants, American as well as foreign, even princes of royal blood, among whom were the Comte de Paris in 1862, and Prince Napoléon Jérome, and Princess Clotide in 1861, visited the Museum with evident satisfaction and pleasure, during what may justly be called its infancy. With Agassiz as a guide, it would have been impossible not to admire both the director and his collections.

After the secession of the Salem party, as the action of the assistant-pupils, in 1864, may be denominated, a better direction of its affairs was soon visible, owing to the continuous and well-directed efforts of Mr. Alexander Agassiz. During the absence of his father in Brazil, he directed the Museum, and although the force employed was much reduced, it worked better, and proved the administrative capacity which has since distinguished Mr. Agassiz.

Agassiz, with his usual generous and enthusiastic encouragement of original observers and students, allowed almost every one the use of the collections; going so far as to send specimens and books to whoever expressed a desire for them. But such a confidence exceeded proper bounds, and as its inevitable result several valuable specimens, books, and papers were lost.

New accessions to the staff of the Museum were made from time to time. Reference to some of the *attachés* has already been made, and it only remains to complete the list, omitting mention of perhaps a dozen, either on account of the small amount of scientific work they have done, or because of their short connection with the Museum.

Mr. J. A. Allen began as a student in 1862, and took charge of the mammalia and birds in 1864, publishing valuable memoirs on the buffaloes, the pinnipeds, etc. He remained in the Museum until 1884, when he accepted a position at the American Museum of natural history in New York.

During the autumn of 1867, Agassiz called from Prussia an entomologist of great reputation, Dr. Hermann Hagen, to take charge of the collections of articulates. His services to the Museum were invaluable. Following the plan adopted by Agassiz, Dr. Hagen succeeded in placing the numerous and rich collections entrusted to his care in fine condition.

In 1869, Dr. G. A. Maack, a pupil of the celebrated professor, Albert Oppel, of the University of Munich, came to the Museum as assistant in charge of the collection of fossil vertebrates. Having passed several years at Buenos Ayres as an assistant in the National Museum, directed by the learned naturalist, Dr. Hermann Burmeister, Maack was well prepared for the work assigned to him in the Agassiz Museum. Unhappily he accepted the position of geologist and naturalist of the United States Darien Expedition sent by the government to explore the isthmus of Panama, Darien, and Choco, and he returned with his health so impaired by constant attacks of Panama fever that his mind became affected; and in a moment of despondency he ended his life. Maack was an excellent observer, a good practical geologist, and well posted in osteology and comparative anatomy, and he sent to the Museum large and valuable collections of mammals, birds, reptiles, fishes, molluscs, and radiates, besides palæontological specimens from Carthagena, the Atrato River, Panama, San Miguel, Cupica Bay, and Napipi River.

Dr. Franz Steindachner, attracted by the magnificent collection of fishes, came to Cambridge from Vienna in May, 1870, to assist Agassiz for two years in their arrangement and determination.   Chosen to accompany Agassiz in his voyage on the *Hassler*, he had the rare privilege of seeing the fish fauna from Cape Cod to the Strait of Magellan, the Galapagos, Panama, Acapulco, San Diego, and San Francisco.   As he had already examined the fauna of the West Coast of Africa, especially at the Cabo de Verde, he had acquired an unusual knowledge of ichthyology.   His work at the Museum was most valuable.   Returning to his native country at the end of 1872, he was placed at the head of the zoölogical division of the Imperial Museum of natural history at Vienna, and has since become direc-tor-general.   Before leaving Cambridge, he said to me : " No naturalist knows fishes like Agassiz; his knowl-edge in ichthyology is unparalleled on account of his researches on both the living and fossil species ! "   An opinion which, coming from such a learned ichthyolo-gist, who was also reserved and careful in his judg-ment, is worth recording.

A conchologist of talent, John G. Anthony, was attached to the Museum, about 1864, to catalogue and label the shells in the Museum. For years he worked faithfully at the task, and, at his death, several years after that of Agassiz, left the conchological collections in an excellent condition.

Leo Lesquereux, during 1868 and 1869, classified the fossil plants at the Museum. After his arrival in America at the end of 1848, Lesquereux studied fossil plants with great success, and had justly become an authority on this subject on both sides of the Atlantic.

I will mention also Charles Hamelin and Messrs. Walter Faxon, Samuel Gorman, and Walter Fewkes, who were attached to the Museum during the last years of Agassiz's life.

Finally Pourtalès resigned his official connection with the Coast Survey in 1873, and took up his residence in Cambridge to assist Agassiz in the general direction of the Museum. As he was the first of the European friends who joined Agassiz at Boston as far back as 1846, it was most appropriate that he should be the last to help him. At the death of his father in 1870, he had inherited a fortune sufficient to place him in an independent position, and he devoted the remainder of his life entirely to his zoölogical studies. Extremely modest and retiring, timid as a child, always a hard worker, but rather slow in all his motions, persistent in his schemes and undertakings, he possessed just the qualities required in the long and weary process of casting and hauling the dredge in the collection of marine animals at great depths. As a curator of the Agassiz Museum, he devoted the last part of his life to the progress of the institution. Frank de Pourtalès was a man always true to his word, a rare and most trusty friend. His original work in natural history was confined to the deep-sea corals of the Gulf Stream area in the Caribbean Sea, and the Gulf of Mexico, and on the Florida coast, to crinoids and to halcyonarians. He died in July, 1889, at the age of fifty-seven years, and is buried in Mount Auburn cemetery, not far from the grave of his old teacher and friend.

A few quotations from the last report of Louis Agassiz and from the first of his son and successor will show the condition and standing of the Museum of Comparative Zoölogy during the last year of the life of its founder and at the moment of his death.

I have heard it said repeatedly, that the organization of the Museum was too comprehensive, that it covered a wider range than was useful in the present state of science among us, and that since it must collapse whenever I should be taken away, it was unwise to support it on so large a scale. The past year has proved beyond question that the Museum is now so organized (vitalized, as it were, with the spirit of thought and connected work) that my presence or absence is of little importance. It will keep on its course without any new or repeated stimulus beyond the necessary appropriations for its maintenance. As to the expense, I cannot feel that it is disproportionate, because when I compare it with that of institutions of the same character I see that they spend much more for smaller results. The only question now is, whether a museum of the first order is needed in Massachusetts, or not. If the Legislature will favour us with a visit, I would gladly submit our institution to the most critical examination of its organization. I think I can satisfy any competent visitor, that by her liberal support of the Museum, our state has earned the right to say, that among civilized com-

munities there is not a purely scientific establishment of higher
character, or distinguished by more active, unremitting original
research in various departments of knowledge. If the same pecuniary support it has had in the last two years can be continued in the
coming years, it will not be long before the scientific world will
acknowledge that the Museum of Comparative Zoölogy in Cambridge has no superior, nay, no equal, in the world (Report of the
Director, in the "Annual Report of the Museum of Comparative
Zoölogy at Harvard College, in Cambridge, for 1872," pp. 4, 5;
Boston, 1873).

Early in 1873 it became apparent that the Museum could not
longer be carried on with the means at the disposal of the Curator.
Repeated assistance from the state and from private sources kept
the institution up to a standard of activity far beyond its own regular resources. As the time drew near when retrenchment seemed
inevitable, Professor Agassiz made an appeal to the Legislature for
support, and with the generosity which has always characterized
their action towards an institution in which the state of Massachusetts has so great an interest, the Legislature appropriated twenty-five
thousand dollars, on condition that a similar sum should be contributed by the friends of the institution towards its support. This
sum was at once subscribed by friends of the Museum, and the
appropriation of the state secured. Soon after this a further sum of
one hundred thousand dollars was presented to the Museum by
Mr. Quincy A. Shaw. These sums gave Professor Agassiz the
means to reorganize the Museum on a very extensive scale. Additional assistants were employed, collections were purchased in
every direction, and a large outlay made to place in safety the
valuable alcoholic collections stored in the cellar of the Museum
building. True to his policy of always using his present means as
a lever for further improvement, nothing was laid up for the future
and by the first of April next the Museum will have to depend
entirely upon its invested funds for its resources. This will entail
a very material reduction in the working force and running expenses, as the regular income of the Museum is somewhat less than

fifteen thousand dollars annually, only half the sum needed to carry on the present scale of operations ("Annual Report of the Museum for 1873," pp. 4, 5 ; Boston, 1874).

The constant success of Agassiz, in obtaining for his Museum appropriations of large sums of money from the Legislature of Massachusetts, is something unique in natural history; for the Museum has been finally turned over by its trustees to Harvard University, a private corporation ; and if Harvard had asked of the Legislature a sum of money, however small, for the foundation of a museum, it would never have been granted. The success is entirely personal, and due wholly to Agassiz's power of persuasion. He quickly became expert in handling the Legislature. When called before the Committee of Appropriations to explain the nature of his wants, he would meet every member of the committee, first in private, then in the committee-room. But before any step was taken, he would call on the Governor, the Lieutenant-Governor, the President of the Senate, the Speaker of the House, the Secretary of the Board of Education, and the Chief Justice of the Supreme Court, all *ex-officio* trustees of the Museum, and consequently in sympathy with its needs. The amount of scientific diplomacy he made use of is something astounding ; for instance, he would detail, with great clearness, the working of the institution, and make it clear that the Museum is an element of education even in the most elementary school of the commonwealth, and that in the future generations there would not be a child who would not have the oppor-

tunity of understanding the scheme of creation as thoroughly as he understood his multiplication table.   He
had the tact to adapt his explanations and his description
of the absolute poverty of the institution, to the listener
and his official position in the state.   Then, after weeks
of such preparatory work at the state house, came the
annual visit of the whole Legislative body, with the
Governor at its head, to the Museum.   Everything was
in readiness for the reception when the six or ten street
cars, filled with legislators, arrived at the University
grounds.   Agassiz conducted them at once into the
various exhibition halls, showing the treasures of each,
and briefly describing the departments.   Afterward, in
the lecture-room, in an informal conversation, he detailed the methods and needs of the institution.   He
always succeeded in winning to his side farmers, tradesmen, and politicians.   After such a visit, the Legislature always voted a new appropriation of public
money; it was only necessary for the President of the
Senate and the Speaker of the House to make speeches
in its favour, and the resolution would easily pass the
three readings without further debate.

Agassiz made stupendous efforts, during the last fourteen years of his life, to obtain seven hundred thousand
dollars to found his Museum.   Less than half of it was
furnished by the state of Massachusetts, and the rest
by private subscriptions, a great part of it coming from
his own family and relatives.   If he had gone to Washington and made only half the exertion he did in Boston, he would have easily obtained from Congress ten
and even twenty millions of dollars to found the United

States National Museum. The University of Cambridge cannot be grateful enough for the service he rendered in identifying himself with it, and founding for its benefit an American institution, which, in many respects, rivals the great museums of London, Paris, Vienna, and Berlin.

CHAPTER XXIII.

1873.

AGASSIZ came back from his journey around South
America with such renewed vigour of mind and body
that he renewed his social duties, and his always hospitable house was often open to his friends and his
large family circle.

At the beginning of 1873 occurred the most extraordinary episode in Agassiz's life. A merchant of New
York, upon seeing in the newspapers the announcement that Agassiz proposed to give a course of instruction in natural history, chiefly designed for
teachers, and students preparing to become teachers,
to be delivered on the island of Nantucket during
the summer months, offered him Penikese Island,
one of the Naushon Islands, in Buzzard's Bay, opposite New Bedford, on the southern coast of Massachusetts, and to complete the gift, an additional
endowment of fifty thousand dollars, for the support

and maintenance of the school. To a man of the optimistic and enthusiastic temperament of Agassiz, the offer was too great a temptation to resist, notwithstanding his age and his broken health. Those near him knew that he was not in a condition to accept such a heavy burden, in addition to the many engagements already assumed. But remonstrances were of no avail; he answered all objections, and after a few weeks of hesitation accepted the gift. As his son says: " It is a new *pompe* added to the many already in activity." Mr. Alexander Agassiz, above all a business man, was justly alarmed at the anticipated expenses of the Museum, without the addition of another burden, the extent of which it was impossible to foresee. Knowing his father's propensity to " faire grand " in everything relating to science, without any regard to expenditure, it is not surprising that he was opposed to the acceptance of the gift. But it was impossible to restrain Agassiz when he had started on any special scheme, and a summer school of natural history had been for years one of his pet desires.

In March, 1873, he wrote to Mr. John Anderson: "It seems to me impossible to do otherwise than accept the great gift you offer. It changes at once an experiment without fixed location or stable foundation into a permanent school for the study of nature, such as the world has not seen before. ... . I am overwhelmed by your generosity [the additional endowment of fifty thousand dollars]. Such a gift, following so close upon the donation of an island, admirably adapted by its position for the purposes of a practical school for natural his-

tory, opens visions before me such as I had never dared
to indulge in connection with this plan."

On the 22d of April, in company with members of
the city government of New Bedford and a number of
invited guests, he visited the island to take formal possession, where the party was cordially welcomed by
Mr. and Mrs. Anderson. Under these circumstances
Agassiz first met Mr. Anderson, and for the first time
saw what was to be one of his future laboratories in
the cause of natural science. The ceremony of the
transfer took place in the house of Mr. Anderson, his
solicitor reading the deed of conveyance.

Agassiz returned from his visit, delighted with the
island and its surroundings. He at once ordered a
building for the laboratory, upon which work was immediately begun, and by the 8th of July the building
was ready for the reception of fifty persons, and the
school of natural history commenced on the appointed
day. The lecture-room was an old barn, and there Agassiz, with bared head, called the pupils to join him in silent
prayer. Whittier's poem on this subject is as follows: —

THE PRAYER OF AGASSIZ.

On the isle of Penikese,
Ringed about by sapphire seas,
Fanned by breezes salt and cool,
Stood the Master with his school.
Over sails that not in vain
Wooed the west-wind's steady strain,
Line of coast that low and far
Stretched its undulating bar,
Wings aslant along the rim
Of the waves they stooped to skim,

Rock and isle and glistening bay,
Fell the beautiful white day.
Said the Master to the youth:

" We have come in search of truth,
Trying with uncertain key
Door by door of mystery;
We are reaching, through His laws,
To the garment-hem of Cause,
Him, the endless, unbegun,
The unnamable, the One
Light of all our light the Source,
Life of life, and Force of force.
As with fingers of the blind,
We are groping here to find
What the hieroglyphics mean
Of the Unseen in the seen,
What the Thought which underlies
Nature's masking and disguise,
What it is that hides beneath
Blight and bloom and birth and death.
By past efforts unavailing,
Doubt and error, loss and failing,
Of our weakness made aware,
On the threshold of our task
Let us light and guidance ask,
Let us pause in silent prayer! "

Then the Master in his place
Bowed his head a little space,
And the leaves by soft airs stirred,
Lapse of wave, and cry of bird
Left the solemn hush unbroken
Of that wordless prayer unspoken,
While its wish. on earth unsaid,
Rose to heaven interpreted.

As in life's best hours we hear
By the spirit's finer ear
His low voice within us, thus
The All-Father heareth us ;
And His holy ear we pain
With our noisy words and vain.
Not for Him our violence
Storming at the gate of sense,
His the primal language, His
The eternal silence !

Even the careless heart was moved,
And the doubting gave assent,
With a gesture reverent,
To the Master well-beloved.
As thin mists are glorified
By the light they cannot hide,
All who gazed upon him saw,
Through its veil of tender awe,
How his face was still uplit
By the old sweet look of it,
Hopeful, trustful, full of cheer,
And the love that casts out fear.
Who the secret may declare
Of that brief, unuttered prayer ?
Did the shade before him come
Of th' inevitable doom,
Of the end of earth so near,
And Eternity's new year ?

In the lap of sheltering seas
Rests the isle of Penikese ;
But the lord of the domain
Comes not to his own again ;
When the eyes that follow fail,
On a vaster sea his sail
Drifts beyond our beck and hail.

Other lips within its bound
Shall the laws of life expound ;
Other eyes from rock and shell
Read the world's old riddles well ;
But when breezes light and bland
Blow from summer's blossomed land,
When the air is glad with wings,
And the blithe song-sparrow sings,
Many an eye with his still face
Shall the living ones displace,
Many an ear the word shall seek
He alone could fitly speak.
And one name forevermore
Shall be uttered o'er and o'er
By the waves that kiss the shore,
By the curlew's whistle sent
Down the cool, sea-scented air ;
In all voices known to her,
Nature owns her worshipper,
Half in triumph, half lament.
Thither Love shall tearful turn,
Friendship pause uncovered there,
And the wisest reverence learn
From the Master's silent prayer.

It was amusing to see Agassiz delivering his lectures, surrounded not only by forty-four students,[1] of both sexes, but by the workmen who were finishing the interior arrangements and erecting a second building. Never had the small island seen so many people collected on its shores. Every one was collecting, examining with microscopes, dissecting, or watching marine animals in

[1] Among the students at Penikese, I will mention only a few who have become celebrated since : Professor C. O. Whitman of Chicago University, D. S. Jordan, President of the Leland Stanford Jr. University (California), Professor William K. Brooks, and Professor Charles S. Minot.

aquaria improvised out of pails and buckets. Agassiz lectured nearly every day, and frequently twice a day, and his passion for teaching had full play. Mr. C. W. Galloupe of Boston made him a donation of his yacht, *Sprite*, and as she was fully equipped, Pourtalès took charge of her and at once began dredging, going out daily, weather permitting, with eight or ten students, and obtaining a variety of specimens which could not be procured from the shore ; and at the close of the school session they went as far as Casco Bay, to dredge for brachiopods and echinoderms that could not be procured in Buzzard's Bay.

Agassiz left Penikese[1] at the end of the summer, when the school broke up, and on invitation of friends visited the mountains for rest, which was an absolute necessity in his present condition of mental and physi-

---

[1] The Anderson School of Natural History at Penikese Island did not survive long after Agassiz's death. The appeal for aid addressed by Mr. Alexander Agassiz to the superintendents of public institutions and presidents of State Boards of Education of the several states, did not find the ready response necessary for the support of the school, and although the expenses were estimated at a minimum, they were too large for the means at the disposal of the director, and the Anderson School was soon a thing of the past. But if its existence was ephemeral, it set a most beneficial example, soon followed by permanent schools of the same sort, created in imitation of the Marine Biological Laboratory of Penikese Island, first, those at Wood's Holl, Mass., one under the direction of the United States Fish Commission, and the other directed by Mr. C. O. Whitman; second, one at Annisquam, and afterward at several other places on the Atlantic and Pacific coasts, under the direction of the Johns Hopkins University, the State University of California, and the Leland Stanford Jr. University, while Mr. Alexander Agassiz, notwithstanding his failure at Penikese in carrying on the school, has since built a fine laboratory at Castle Hill, near his summer residence at Newport, Rhode Island, where researches on living marine animals are made every summer under his direction and at his expense.

cal fatigue.   The effort had been too great, and the
strain upon his health beyond reason.   When among
pupils it was impossible to restrain him.   He must
teach.   Teaching was as natural to him as breathing to others; but after his illness of 1870, he was
obliged to exert himself to deliver his lectures, and
it was often painful to see him forcing his voice
through his over-fatigued throat.   His throat was the
weak point in his herculean frame.   However, October,
1873, found him again at his post in his Museum, and
he began a course of lectures on the radiates from
their first appearance until the present time.   At the
same time he dictated to Mrs. Agassiz an article for
the "Atlantic Monthly," on "Evolution and Permanence of Type," which did not appear until January,
1874, after his death.   As it is his last production, it
may be taken as "Louis Agassiz's Scientific Will"; and
a few quotations will serve to show his strong convictions on the most exciting of all natural history subjects.

The law of evolution, so far as its working is understood, is a
law controlling development and keeping types within appointed
cycles of growth, which revolve forever upon themselves, returning
at appointed intervals to the same starting-point, and repeating
through a succession of phases the same course.   These cycles
have never been known to oscillate or to pass into each other;
indeed, the only structural differences known between individuals
of the same stock are monstrosities or peculiarities pertaining to
sex, and the latter are as abiding and permanent as the type itself.
Taken together the relations of sex constitute one of the most
obscure and wonderful features of the whole organic world, all the
more impressive for its universality. . . .

Under the recent and novel application of the terms "evolution"
and "evolutionists," we are in danger of forgetting the only process

of the kind in the growth of animals which has actually been demonstrated, as well as the men to whom we owe that demonstration. Indeed, the science of zoölogy, including everything pertaining to the past and present life and history of animals, has furnished, since the beginning of the nineteenth century, an amount of startling and exciting information in which men have lost sight of the old landmarks. In the present ferment of theories respecting the relations of animals to one another, their origin, growth, and diversity, those broader principles of our science — upon which the whole animal kingdom has been divided into a few grand comprehensive types, each one a structural unit in itself — are completely overlooked. . . .

The time has, perhaps, not come for an impartial appreciation of the views of Darwin, and the task is the more difficult because it involves an equally impartial review of the modifications his theory has undergone at the hands of his followers. The aim of his first work on " The Origin of Species " was to show that neither vegetation nor animal forms are so distinct from one another or so independent in their origin and structural relations as most naturalists believed. This idea was not new. Under different aspects it has been urged repeatedly for more than a century by de Maillet, by Lamarck, by E. Geoffroy Saint-Hilaire and others ; nor was it wholly original even with them, for the study of the relations of animals and plants has at all times been one of the principal aims of all the more advanced students of natural history ; they have differed only in their methods and appreciations. But Darwin has placed the subject on a different basis from that of all his predecessors, and has brought to the discussion a vast amount of well-arranged information, a convincing cogency of argument, and a captivating charm of presentation. His doctrine appealed the more powerfully to the scientific world because he maintained it at first not upon metaphysical ground, but upon observation. Indeed, it might be said that he treated his subject according to the best scientific methods, had he not frequently overstepped the boundaries of actual knowledge and allowed his imagination to supply the links which science does not furnish. . . .

The excitement produced by the publication of " The Origin of Species " may be fairly compared to that which followed the appearance of

Oken's "Natur-Philosophie," over fifty years ago, in which he claimed that the key had been found to the whole system of organic life. . . .

Darwin's watchwords, "natural selection," "struggle for existence," "survival of the fittest," are equally familiar to those who do, and to those who do not, understand them; as well known, indeed, to the amateur in science as to the professional naturalist. His theory is supported by a startling array of facts respecting the changes animals undergo under domestication. . . .

The final conclusion of the author is summed up in his theory of Pangenesis. And yet this book does but prove more conclusively what was already known; namely, that all domesticated animals and cultivated plants are traceable to distinct species, and that the domesticated pigeons, which furnish so large a portion of the illustrations, are, notwithstanding their great diversity under special treatment, no exception to this rule. The truth is, our domesticated animals, with all their breeds and varieties, have never been traced back to anything but their own species, nor have artificial varieties, so far as we know, failed to revert to the wild stock when left to themselves. Darwin's works and those of his followers have added nothing new to our previous knowledge concerning the origin of man and his associates in domestic life, the horse, the cow, the sheep, the dog, or, indeed, of any animal. The facts upon which Darwin, Wallace, Haeckel, and others base their views are in the possession of every well-educated naturalist. It is only a question of interpretation, not of discovery of new and unlooked-for information. . . .

It has even been said that I have myself furnished the strongest evidence of the transmutation theory. This might, perhaps, be so, did these types follow, instead of preceding, the lower fishes. But the whole history of geological succession shows us that the lowest in structure is by no means necessarily the earliest in time, either in the vertebrate type or any other. Synthetic and prophetic types have accompanied the introduction of all the primary divisions of the animal kingdom. With these may be found what I have called embryonic types, which never rise, even in their adult state, above those conditions which in higher structures are but the prelude to the adult state. It may, therefore, truly be said that a great variety of types has existed from the beginning. . . .

The world has arisen in some way or other.   How it originated is the great question, and Darwin's theory, like all other attempts to explain the origin of life, is thus far merely conjectural.   I believe he has not even made the best conjecture possible in the present state of our knowledge. . . .

I would add as a *résumé:* Man has not yet been able to create, or "evolve," if the word is more acceptable to the followers of Darwin's theory, a single true species of animal or plant; but *per contra* he has certainly the power to destroy them, several species of animals having been exterminated during the last two centuries by men — not one of whom knew anything about the origin of species, according to Darwin, Lamarck, Haeckel, or Huxley.   Destruction is certainly easier than evolution.

The last, but not the least, natural history surprise enjoyed by Agassiz came from Newfoundland.   Fishermen in Conception Bay, in a battle against a gigantic squid, succeeded in cutting off and securing an arm of the beast nineteen feet long.   The body of the animal was sixty feet long, and his diameter not less than five feet.   The state geologist of Newfoundland, Mr. A. Murray, wrote me a long letter[1] on this remarkable monster, which I hastened to communicate to Agassiz. The following is Agassiz's letter to Mr. Murray on the subject: —

CAMBRIDGE, MASS., Nov. 25, 1873.

My dear Sir, — My friend Marcou has communicated to me your most interesting letter; and I am delighted at last to have so direct information concerning the gigantic cephalopods of the Atlantic, of which so much has been said since the days of Pontoppidan in his

[1] This letter from Mr. Murray was published in "The American Naturalist," Vol. VIII., pp. 120–123.   February, 1874.   Salem.

"Norwegian Fables of the Kraken."   I will now hunt up everything
that is worth noticing upon the subject; and if you will allow me
an examination of your specimen, the zoölogical characters of the
creature might be made out from the part preserved, as we do of
imperfect fossil remains.   I would also ask leave to publish the
substance of your letter to Mr. Marcou, in connection with this.

> With great regard, yours very truly,
>
> Ls. Agassiz.

Alex. Murray, Esq.,
St. John's, Newfoundland.

This letter and the two following, of which a fac-
simile is here given, were the last scientific letters
written by their illustrious and lamented author, the
last one on the 26th of November, 1873.

Museum of Comparative Zoology,

CAMBRIDGE, MASS. *Nov. 25.*
1873

Mon cher Marcou,

Merci pour la lettre et la photographie
que Mr. Murray vous a adressée. C'est fort curieux
et avec votre permission j'en publierai le contenu accompagné
de remarques, si Mr. M. veut m'envoyer une des grandes
ventouses pour la comparer à celles des espèces de Céphalopodes
connues sur nos côtes. Je vais lui écrire aujourd'hui
même ce but.

Tout à vous

L. Agassiz

Museum of Comparative Zoology,

CAMBRIDGE, MASS.

[handwritten letter in French]

Mon cher Marcou,

J'ai fait copier la lettre de M. Murray et je vous retourne l'original. Plus je considère cette trouvaille et plus elle m'intéresse. C'est vraiment important pour l'histoire des Céphalopodes.

Tout à vous

F. Agassiz

(Post-marked Nov. 26, 1873.)

On the 2d of December Agassiz delivered his last lecture before the Massachusetts Board of Agriculture at Fitchburg, on "The Structural Growth of Domesticated Animals." On the 5th he enjoyed, as usual, his weekly family dinner, with all his children around him, smoked cigars, contrary to the special order of Dr. Brown-Séquard; but the next morning, the 6th, he complained of a dimness of sight, of feeling "strangely asleep," and of great weariness. He went, nevertheless, to his Museum, but soon returned, and lay down in his room. It was his last illness. Paralysis of the larynx rapidly developed; and all the care and skill of Dr. Brown-Séquard, then in New York, who came at once to the side of his friend, and of another friend, Dr. Morrill Wyman, could not stay the mortal disease.

Agassiz had been in great dread of softening of the brain, of which his friend, Professor Bache, had died in 1866, after a very long and most painful illness. He often expressed the hope that he should disappear suddenly; and his wish was in great part realized, for he lingered only eight days. It was, however, hard for him to die just when fortune had at last smiled on him and all his children; and when everything was ready for the realization of the two dreams of his life, — a great museum and a practical school of zoölogy; but the old Arab proverb proved true also for him: "When the house is ready, death walks in."

He had so many schemes, and was so full of projects, that desire to prolong life was still very strong in him, even after he was stricken by such a grave illness. The presence at his bedside of the great physiologist, Dr. Brown-Séquard, encouraged him, and it was not until the last day that he gave up all hope. During his short illness, which was undisturbed by acute suffering, he received every comfort which his family could divine.

Agassiz resumed his native language as soon as Dr. Brown-Séquard came, and used it until the end. When all hope of recovery was given up, during the last eighteen hours, he often said, "Tout est fini!" And when the last moments came, all retired to the adjoining room to let him finish his life in complete quietness;

Grave of Louis Agassiz at Mount Auburn (Front).

while they kept watch over him from the open door, relieving one another from time to time. It was Pourtalès who, at the last moment, was surprised to see him rise in his bed, and to hear him exclaim, with great distinctness, "Le jeu est fini!"[1] Then he fell back, and died, shortly after ten o'clock P.M., the 14th of December, 1873. Life for him had been a long and successful play, well filled from beginning to end. A post-mortem examination was made by Drs. Brown-Séquard, Jeffries and Morrill Wyman, assisted by five other physicians. The brain was found to be very large and heavy, like that of George Cuvier, and traces of disease were recorded for a period dating back at least twelve years.

The funeral took place on the 18th, at 2 P.M., in Appleton Chapel, in the College ground, Harvard Square. Rev. Dr. Andrew P. Peabody, professor in the College, conducted the service, according to the King's Chapel liturgy, of Boston. It was simple, all ceremonies except the strictly religious rites being dispensed with. The church was crowded with the most noted assembly ever seen in New England, including the Vice-President of the United States, the Governor, Ex-Governor, admirals, major-generals, poets, naturalists, savants, and distinguished ladies, together with the little band of Europeans who came with Agassiz to the New World, and all the members of the faculty of the University, with the students in a body.

It was a winter afternoon, without snow, and not a

---

[1] This recalls the exclamation of Rabelais at the moment of his death, "La farce est jouée."

cold day. When the benediction was pronounced, the body was removed, the organ playing the " Dead March in Saul." A long procession followed to Mount Auburn, where the remains were buried in one of the most beautiful parts of the cemetery, very near the grave of Agassiz's friend, Felton. The monument erected over his grave is symbolic of one of his most remarkable discoveries. It is simply an Alpine boulder, weighing twenty-five hundred pounds, from the moraine of the glacier of the Aar. This granite block was selected by his cousin, M. Auguste Mayor, of Neuchâtel, from the lateral moraine, not far from the spot where the celebrated " Hôtel des Neuchâtelois" once stood. It was carried with great difficulty, "à force de bras," from the glacier to the Bernese Oberland village of Imhof, a distance of twenty-five miles, thence on a wagon to the railroad station of Thun. Around this superb boulder — on which are engraved on one side the words, " Jean Louis Rodolphe Agassiz," and on the other, " Born at Motier, Switzerland, May 26, 1807; died at Cambridge, Mass., December 14, 1873 " — there are several pine trees which formerly grew near the celebrated boulder called " Pierre-à-Bot," above the city of Neuchâtel, and which were successfully transplanted, and now shelter the Agassiz lot.

Grave of Louis Agassiz at Mount Auburn (Back).

CHAPTER XXIV.

DR. PROFESSOR HERMANN LEBERT, the anatomist and naturalist, says: "Agassiz was one of the most brilliant men of his time. Young, handsome, of an athletic constitution, gifted with a captivating eloquence, his spirit was animated by an insatiable curiosity, his memory excellent, his perspicacity rare and very keen, and his way of judging and coördinating facts highly philosophical in its tendency" ("Actes de la Société helvétique des sciences naturelles réunies à Bex," août, 1877, p. 149). No one was better fitted to give an exact description of the physical and scientific characteristics of Agassiz than Dr. Lebert, who had known him well during the twelve years of his greatest scientific activity, from 1834 to 1846.

Agassiz was a little above the average height, although not tall. He was squarely built, with broad shoulders and a powerful and well-proportioned body, and with remarkably large, and at the same time well-formed, hands, which he always used most skilfully. They were the hands of an artist or of a naturalist, ready to use the pencil, the hammer, the scalpel, or

217

the microscope, and his manner of shaking hands was very cordial and friendly. He stood firmly, though his feet were rather small in comparison with his herculean structure, and seemed formed for walking; indeed, he was all his life a capital pedestrian, both on level ground and among the Alpine mountains.

His head was simply magnificent, his forehead large and well developed; and his brilliant, intelligent, and searching eyes can be best described by the word fascinating, while his mouth and somewhat voluptuous lips were expressive, and in perfect harmony with an aquiline nose and well-shaped chin. His hair was chestnut colour and rather thin, especially on the top of his head; indeed, after he was thirty-six years old he showed signs of baldness, which greatly increased after his fiftieth year. The only part of Agassiz's body which was not in harmony with the rest was his short neck, which gave him the appearance of carrying his head on his shoulders, — a defect which he possessed in common with Napoleon Bonaparte. It was his weak point, and the part which failed first.

He was easily moved to tears, and at times cried like a child. He had spells of laughing, which sometimes seemed forced, but which were perfectly spontaneous. It was almost impossible for him to conceal his emotions. This remark applies more especially to the first forty years of his life; later, he was less apt to show his feelings. During the first half of his life he was seldom angry, however great the provocation might be. But after 1853, he quite often got into a passion, even losing control of his words, although he

never ceased to be a gentleman, and was careful
not to wound too deeply his adversaries or the person with whom he was discussing. As one of his
most constant and bitter enemies said, he was "un
bon enfant." During his youth and student life, and
even as late as his "séjours" at the "Hôtel des Neuchâtelois," he used to sing, and even to yodel, like the
Tyrolese peasants and the Oberland guides; but he
completely ceased as soon as he landed on American
soil.

Fine clothes never attracted him. He was, on the
contrary, rather inclined to wear the most common and
unbecoming suits and a slouch hat during winter and
summer, and I do not believe that during his whole
stay in America he ever wore a silk hat. At Neuchâtel, his dress was most ordinary, notwithstanding
the rather formal society in which he moved. But in
Paris, during his long visit of 1846, he was obliged to
follow the customs of other savants, conformed to ever
since the time of Cuvier and Humboldt, viz. a black frock
coat, white cravat, and high hat. Alexander von Humboldt affected to wear such ceremonial dress even when
he explored the Ural Mountains and Central Asia, and
Leopold von Buch, as well as Élie de Beaumont, did
the same during their geological excursions. It was
not becoming to Agassiz, however, and he was delighted
when he arrived in America to find that every one
dressed as he pleased, without any ceremony or con-
vention of any sort. He very seldom wore gloves,
never carried a cane, except an alpenstock, and very seldom used an umbrella. When in Neuchâtel at official

meetings, he wore over his coat the ribbon and cross
of the Red Eagle of Prussia, but after leaving Neuchâtel he wore no decorative ribbon of any kind, notwithstanding that he possessed that of knight and
officer of the Legion of Honour of France, besides the
Prussian order. On the whole, Agassiz was extremely
simple, and did not like to make an appearance different from that of ordinary people in his neighbourhood.

Being a good pedestrian, he carried on his back in
all his excursions in Switzerland and Southern Germany a knapsack of the kind now called " sacs de
tourists." At Neuchâtel one was always lying in the
library at the foot of his desk, with hammers and
papers and specimens. He did not bring one with
him to America, using a carpet-bag instead.

In religion, Agassiz was very liberal and tolerant,
and respected the views and convictions of every one.
He was opposed to all form and exaggeration, and
did not like theology, but avoided, as far as possible,
all discussion of the subject. He was neither a sceptic
nor a scoffer, and Dr. Karl Vogt, with his unceasing
sceptical and cynical attacks against the Bible, shocked
him so much that, notwithstanding Vogt's great talents
as a naturalist, Agassiz was glad when he left Neuchâtel.

In his youth and early manhood, Agassiz was undoubtedly a materialist, or, more exactly, a sceptic;
but in time, and little by little, his studies led him to
a belief in a divine Creative Power. He was more
in sympathy with Unitarianism than any other Christian denomination. He was married to his second wife

by a Unitarian minister, and his funeral service was conducted by another, both named Peabody, but not related. Agassiz always avoided the society of ministers and clergymen in general, because, as he said, "he saw too many of them in his youth."

In society, he was fond of meeting rich and influential men, being always desirous of acquaintance with those possessing a great name and in high social or official positions. He was an aristocrat by nature, with a strong mixture of popular and democratic habits and manners, and always ready to distribute large doses of "l'eau bénite des cours," in which connection he had been at a good school, with Alexander von Humboldt, who was a perfect courtier.

He was not fond of the society of other savants, often having occasion to complain of their conceit and curiosity. He did not know how to repulse the too familiar manners, and even insolence, to which celebrated men are often exposed. Poets, litterati, historians, philologists, lawyers, and especially merchant princes, were more to his taste. Military officers had no attraction for him; but he was very fond of naval officers and captains of merchant steamship lines.

Agassiz, who was a brilliant conversationist, had also the most winning manners; and he easily induced those around him to accept and even to share in his enthusiasm. If any one resisted, he was not discouraged, and displayed a true coquetry in his efforts to conquer. Easily approached, he met you most openly, with expressions of frankness, mingled with agreeable surprises, and brought you to his view or side almost before you were aware of it.   However, Agassiz seldom revealed himself entirely, and after his fiftieth year he never did so.   He possessed the shrewd simplicity characteristic of the Vaudois peasant, and went willingly beyond the mark in order to discover the true meaning of your thought.   At first his design did not appear; and you were led to look upon him as a man who said all and even more than he knew and thought.   But soon you were obliged to admit that your first impression of the man was erroneous; you found an unconquerable opposition to anything of which he disapproved. Agassiz was very ready to make promises; he asked favours in every direction, and then he was apt to forget the conditions under which they were granted. He was at the same time a dreamer and a man of action, dreaming aloud, and taking the public as a confidant of everything which came into his mind. He was not to be taken at his own word, but it was necessary to allow a large margin for contingencies and changes.   When surrounded by material difficulties, he fortified his spirit by a marvellous power of always hoping for better times, having an absolutely unshaken confidence in himself.

He was one of those whose hands and heart are always open; for whom work is the main path of life, and, at the same time, a great pleasure and not to be interfered with, — a sort of prodigal child.   "Il ne vivait que pour les autres, s'il n'y avait eu personne pour le regarder tout le temps, il n'aurait pas existé," — a sentence of a distinguished French author, which applies fully to Agassiz.   He always thought that he

had not enough friends or assistants around him. His
home was a sort of "phalanstery" of savants. He
was always inclined to trust too much to his persuasive-
ness; and if it often helped him, sometimes it acted
otherwise, and was the cause of very regrettable mis-
takes. He would take for genuine any compliments
which flattered some hobby or weak point, and no other
explanation can be given of some of his blunders. He
repeatedly let escape him — at least in America — several very able savants, retaining instead persons whom
he very well knew were not so capable.

I have already had occasion to notice his generosity.
Agassiz was a kind-hearted man; he helped many with
money, and always in an unostentatious way. It is true
that he was often deeply in debt; but as soon as he
had money he distributed it almost lavishly, without
thinking of the morrow. One anecdote will show how
full of liberality and charity he was. A Neuchâtel
merchant, established at New York, had suffered
pecuniary misfortunes, and become a pauper, perhaps
on account of his bad habits. This man had absolutely no claim on Agassiz, who came to Neuchâtel
several years after he had left there; but he had
kindly received the artist, Burkhardt, at New York,
in 1844, and this was reason enough for Agassiz,
who ordered the agent of his cousin, M. Auguste
Mayor, to pay fortnightly to this poor Neuchâtelois
an allowance sufficient for his support until he died,
several years after. This occurred during and at the
end of the Civil War, before any of Agassiz's children
became wealthy.

Mrs. Agassiz says: "The ability, so eminently possessed by Agassiz, of dealing with a number of subjects at once, was due to no superficial versatility.   To
him his work had but one meaning.   It was never dis-
connected in his thought, and therefore he turned from
his glaciers to his fossils, and from the fossils to the living world, with the feeling that he was always dealing
with kindred problems, bound together by the same
laws."[1]   And she adds that Agassiz followed all his
life a unity of plan in his scientific researches.

Professor Karl Vogt,[2] of Geneva, who was associated
with Agassiz for five years of the most active and scientifically productive part of his life, says: "I never
met with another man gifted with such remarkable
talent in the zoölogical domain.   Agassiz, better than
anybody else, made discoveries in collecting materials
and looking rapidly over collections; but after a first
hasty examination and classification, and when it was
time to study methodically the specimens, then he
escaped and shut himself up like the folding of the
blades of a pocket-knife, and it was most difficult to
bring him back to the work only sketched out. . . .
He was wanting in character, being like a piece of
wax, which retains the mark of the last hand that has
held it, and like a weathercock, which turns all around

[1] "Louis Agassiz," by Mrs. E. C. Agassiz, Vol. I., p. 336.
[2] Karl, or Carl, or Charles Vogt, born at Giessen, the 5th July, 1817,
died at Geneva, the 5th May, 1895.   He was the last surviving member
of the scientific household of Agassiz at Neuchâtel.   Dr. Charles Frédéric
Girard, another assistant, preceded him by only a few months, having died
on the 29th of January, 1895, at Levallois-Perret, a suburb of Paris, at the
age of seventy-two.

the horizon, believing that it has remained motionless, pointing all the time in the same direction." [1]

These two diametrically opposed views are both exaggerated. The truth lies between them. Agassiz was capricious in the extreme, very versatile, attracted easily by any new object or subject; and he had the faculty of almost completely forgetting works half done or only sketched. He lacked persistence and steadiness at work requiring long and difficult observations. Like a splendid butterfly, he flew from one delight to another. But to compare Agassiz to a weathercock or a piece of wax is a great mistake. When his opinion was formed on a subject, it was impossible to move him. He would listen to all the objections, assent more or less to what was said, but in the end do only what he wanted to do. That something sternly practical mingled with Agassiz's habitual idealism was well proved by his Museum. He did not carry it out entirely, as he proposed to do at the start; but had he lived twenty years longer, his ideal Museum would have become a reality.

He began and abandoned successively many subjects. For instance, after the publication of his "Poissons fossiles" and the "Poissons du vieux grès rouges ou système Dévonien"—that is to say, after 1844—he never took up the subject again, with the single exception of the study of a few fossil teeth, collected in California, and described in Vol. V. of the "Pacific Railroad Explorations"; (Washington, 1856). He never returned to the glacier of the Aar after his hurried visit in 1845. Fossiles echinoderms were also

[1] "Eduard Desor, Lebensbild eines Naturforschers," p. 18.

a favourite study with him from 1833 to 1846, when
he wholly abandoned them.   The same is true of the
Mya and *Trigonia*.   He announced ten volumes of his
"Contributions to the Natural History of the United
States," and only four were published; and, but for his
son, his researches on the Florida corals would never
have been issued.   Part II. of his "Principles of
Zoölogy" was never published.

Notwithstanding these serious defects, it is impossible
not to admire his great scientific intelligence, and not to
recognize his immense scientific force.   No one was
such an able instigator of scientific researches.   He
had a magnetic power, and he used it constantly, whatever the subject to be investigated might be.   His two
principal passions in natural history were teaching and
collecting specimens.   As a teacher he was unrivalled
and unique; and from the first, as a student at Zürich
until the last ten days of his life, we may say that he
taught.   He was always ready to deliver a lecture, —
on the glacier of the Aar, at the Little Academy at
Munich, at Neuchâtel, at Boston, at Rio Janeiro, at
Lake Superior, at sea, on the Amazons, at Cambridge,
at Penikese Island, anywhere.   He would withdraw for
half an hour at most, sometimes for only ten minutes,
and then would begin on the subject chosen, speaking with an abundance of detail, broad general views,
and philosophical conclusions.

As to his other passion, that of collecting specimens
and organizing museums, he was a man of wonderful
resource.   He was insatiable, and had a real mania for
possessing and keeping everything; and he never re-
jected any specimen or drawing.

An anecdote will show how persistent and skilful he was, when he wanted a rare and valuable specimen. Among the few specimens I had kept from my numerous geological explorations was the head of a mammifer of the Miocene from Nebraska, showing the brain, with even a little reddish colour of the animal's blood on it. Agassiz tried two or three times to get it for his Museum. I resisted, wishing to keep it as a memento of my excursion in Nebraska in 1863. When on the point of leaving Cambridge for a long sojourn in Europe in 1864, Agassiz gave a large dinner party in my honour; and as soon as we were all seated at table, in a loud voice, with an imploring tone and in the most friendly way, he begged for that specimen so hard that it would have seemed cruel to deny his request. In fact, that day he acted like a spoiled child who wanted a long-desired toy. Of course he got it.

Agassiz's first plan was to work at living and fossil fishes, — an immense domain for a naturalist. He added to it the fossil echinoderms and afterward the living echinoderms. All the rest of his work came accidentally or incidentally, but not as a result of a unity of plan. He studied glaciers as a pastime, and to prove that the theory of de Charpentier and Venetz was wrong. His researches on the *Mya* and *Trigonia* were prompted by Gressly's discoveries of fine and rare specimens of these two families of molluscs. As soon as he arrived in America, he turned to turtles and jelly-fishes, and began to work on them with the help, first of Charles Girard and Desor, afterward of Mills and Clark. Agassiz was too easily drawn from one study to another.

He was absolutely devoid of the business faculty, — a defect which would have been of little consequence, if he had not always engaged in undertakings involving great expense and requiring financial capacity of no small order.

Like his master, George Cuvier, Louis Agassiz's personality has strongly marked the natural history of the middle of the nineteenth century. With certain similarities, they present a great many more contrasts, due mainly to the differences in the time of their existence, and also to their peculiar temperaments. Both born at the foot of the Jura Mountains, one at the north, at Montbéliard (Doubs), the other at the east, at Motier (Fribourg), they were descended from families essentially Jurassic, Protestant, and including Protestant ministers. Cuvier came originally from the small village of Cuvier, near Censeau, in the department of Jura, while Agassiz came from the small town of Orbe and the village of Bavois in the Jura Vaudois, at a distance of only twenty-five miles from Cuvier. Their youth was passed in countries using the German language; Cuvier's at Stuttgart, and Agassiz's at Bienne, Zurich, Heidelberg, and Munich. On both, German education left indelible marks, more especially during the first half of their scientific life.

Cuvier was, above all, a careful observer of facts. In all zoölogical questions he was never led by his imagination; the only field in which he followed preconceived theory was in general geology, in what he has called the "Révolutions du Globe." On the contrary, Agassiz was inclined to theorize; his brilliant imagina-

tion was constantly alert and given to prophesying the
future of science. However, having seen much more
of the world than Cuvier ever did, his large practical
experience often put a limit to his audacious generalizations, bringing him to more just and rational ideas,
more especially in regard to physical geography.

Cuvier was an excellent practical geologist and observer in the field, and he understood, and may be said
to have created, all the principles of stratigraphy and
of classification of strata.   However, he failed completely in trying to maintain the question of the universal deluge, and the biblical genesis, notwithstanding
many contradictory facts well known to him, and which
he systematically ignored ; as witness his celebrated
command to his assistant, Laurillard, to throw out of
his laboratory window the skeleton of the fossil man
of Lahr (Grand Duchy of Baden), found in the loess
(Quaternary) by Ami Boué, saying : "cela vient d'un
cimetière."   Cuvier thought that no human bones could
be fossil remains, an opinion often disproved by facts
since 1829.   Boué, justly wounded by this rash exclamation of Cuvier, calls it his " hypocrisie biblique," — a
phrase which he extends even to Agassiz in his "Auto-
biographie," 1879, p. xix. of the catalogue of his works.
Cuvier and Agassiz were unwilling to mix science and
religion, and from their education and their connection
with Protestantism, did not feel justified in accepting
facts which seemed probable only, but which lacked
substantial and repeated proofs.   Neither of them was
hypocritical, having too great a respect for science to
merit such a grave condemnation.   Boué went too

far in his criticism,[1] although his discovery of a fossil Quaternary man is a fact now fully accepted.

Agassiz was not a good practical geologist, like Cuvier. His active spirit did not allow him to follow patiently the always long, tedious, and often too-fatiguing researches of practical geology. He wanted the results which he could promptly obtain in the drawers, on the shelves, and in the glass cases of large collections. There Agassiz had not his equal, being even quicker than Cuvier.

Both spoke slowly and with that drawling accent peculiar to the Jurassic peoples, and at their first meeting they recognized one another as children of the Jura. Agassiz kept the accent almost to his sixtieth year; as for Cuvier, he kept it until the end of his life, in consequence of his daily intercourse with his countrymen from Montbéliard, his brother Frédéric, his assistant, Laurillard, and his pupil and cousin, the naturalist Duvernoy.

Cuvier was very grave, while Agassiz, on the contrary, was always laughing, or, at least, smiling. Cuvier had a special aptitude for all kinds of knowledge, and possessed talents to fill any official position, such as professor, general inspector of public instruction, state councillor, great chancellor of the University, or secre-

---

[1] Here is the quotation regarding Agassiz : " Parmi les savants de renom Cuvier n'est pas du reste le seul, qui ait préferé l'hypocrisie à la vérité, l'exemple le plus connu est celui d'Agassiz, qui pour s'assurer sa position en Massachusetts et y pouvoir établir un superbe musée zoologique à Cambridge passa sous les fourches caudines du protestantisme méthodiste le plus absurde." This is a remarkable example of misrepresentation and misunderstanding of the true position of Agassiz in Massachusetts.

tary of public instruction, peer of France, perpetual
secretary of the Academy of Science, etc., etc., while
Agassiz limited himself all his life entirely and exclus-
ively to natural history.   Both possessed an extraordinary memory, and both were remarkably gifted with the
faculty of order; both were capable of long labour, and
at the same time both worked with great facility.   With
them work was always easy.   They did it without effort;
it was natural to them.   But neither was inventive;
both saw facts and observed them sharply, but neither
thought to link them by theories calculated to conduct
to the discovery of other facts.   They were " terre
à terre" naturalists, while Lamarck, Geoffroy Saint-
Hilaire, Darwin, Huxley, looked forward to the future,
prophesying, and always ready to call to their help
suppositions and probabilities.

Physically, Cuvier and Agassiz resembled each other
in possessing enormous heads and largely developed
brains, while neither Lamarck nor Darwin were abnormal as regards size and development of the head.
In a crowd Cuvier and Agassiz always attracted attention, and were distinguished at once as uncommonly
fine-looking men, while Lamarck, Darwin, and Huxley
passed unnoticed.

Agassiz did not possess the original ideas, or the
great sagacity, or the depth of view of Cuvier.   He did
not open new roads to natural history, but he enlarged
greatly all those which were pointed out by others.   If
Cuvier had an enormous influence on the future of
science and on the savants themselves, Agassiz had
more influence on the masses; he made science more

popular, gave a strong impulse to the development of questions very little known before him, and created a more elementary method of teaching. Agassiz delighted in making pupils, and was always on the lookout for applause from all his hearers, whoever they might be, savants or populace. Cuvier, on the contrary, never took the trouble to make pupils, although he left several after him, among them Agassiz and Richard Owen; he never courted applause nor popularity. Cuvier took care to screen himself, and preferred the solitude of his laboratory and library, while for Agassiz solitude was insupportable; he wanted to be surrounded at all times by pupils or admirers. He courted bustle. This is a very unusual characteristic among savants, who are generally more or less retiring, and conduct their researches in the solitude of a laboratory, far from all distractions. As soon as Agassiz had found something new, he proclaimed it even before he had obtained all the proofs. He was always anxious to make an impression on his surroundings and his contemporaries. He was a leader of men, and above all a charmer. Cuvier, on the contrary, was difficult to reach, always on his guard, and very reserved. He did not care about publicity, but he was extremely desirous to make discoveries and keep them secret, until he had deduced all the consequences, and proved them beyond question.

If Cuvier showed great superiority and inventive genius in his classification of the animal kingdom, in his comparative anatomy, his restoration of the forms of fossil vertebrates, his description of the geology of

Paris basin in collaboration with Alexander Brongniart, and in his celebrated lectures at the Jardin des Plantes, and at the College de France, Agassiz rose very high in his study of the " Poissons fossiles," the living fishes, the echinoderms, the *Myae*, the embryology of the turtle, the Acalephs, in the description of the glacial epoch in collaboration with Venetz and de Charpentier, and in his popularization of natural history in North and South America, and finally in his creation of a great museum at Cambridge, and of a great marine biological laboratory at Penikese.   Both were creators, each in his own way.   From 1795 to 1873 these two savants " de très grande envergure " gave to natural history the most important impulse which it has ever received, divulging facts more numerous and more clearly founded on exact principles than any other naturalists who preceded them.   If Cuvier was superior to Agassiz as a classifier, and a creator of several parts of natural history, Agassiz was above Cuvier as a lover of nature, and a popularizer of science.   No naturalist has admired every object of natural history with the enthusiasm of Agassiz.   He stood in ecstasy before a zoölogical specimen; whether it was living or fossil was of no importance to him.   I doubt if any one has ever handled a specimen with such reverence and veneration as Agassiz always did.   Cuvier will always occupy a very exalted position in natural history.   He is above the rank and file; while Agassiz is only in the first rank of Cuvier's pupils.   Agassiz is a brilliant satellite who has moved in the orbit traced by Cuvier; but what an orbit! and what a brilliant light!

APPENDIX.

APPENDIX A.

———•◦•———

BIOGRAPHIES OF LOUIS AGASSIZ.

THE biographies of Louis Agassiz are numerous. Many are mere sketches, and more are only repetitions without any original facts. Scientific periodicals, and even literary reviews and political newspapers, have published a number of articles on Agassiz. I shall quote only those containing original matter. The biographical sketches in dictionaries and encyclopædias, such as the "Dictionnaire des Contemporains," by Vapereau; Appleton's "Cyclopedia of American Biographies," and others in English, French, and German, are all compilations, more or less well executed, without real value, except to popularize his name.

Several papers and books have been quoted, which contain original and important facts bearing on the life of Agassiz, although they were published for other purposes, and under titles which do not indicate Agassiz's connection with the subject treated.

I. DURING HIS LIFE.

Only three original biographies were published during his life.

1845. — The first appeared at Geneva in 1845–47, in the "Album de la Suisse Romane," Vol. V., p. 1, 4to, with portrait. The title is *Agassiz*. The name of the author is not given, but in the Table of Contents, at the end of the volume, it is entered as "Par F. J. Pictet," the celebrated naturalist of Geneva. As I have mentioned, the manuscript was sent to Agassiz, at Neuchâtel, for correction.

This article did not make its appearance until Agassiz was already in America; and it is doubtful whether Agassiz ever received a copy of it, or ever saw it in print. It is excellent, and, although short, is the best biographical sketch we have, so far as it goes; that is to say, up to 1845, when Agassiz was on the point of leaving Neuchâtel for America.

By virtue of his own studies of fossil fishes and fossil invertebrates, Pictet was able to make a just estimation of the merit and originality of Agassiz's researches; and his liberty of judgment and high sense of justice enabled him to examine, without prejudice, the *rôles* played in the glacial question by Venetz, de Charpentier, and Agassiz.

1847. — *The Life and Writings of Agassiz*, published in Boston, December, 1847, in the "Massachusetts Quarterly," Vol. I., pp. 96–119. The author, Mr. J. Elliot Cabot, whose name is not, however, attached to the article, wrote it from materials furnished by Agassiz's secretary, M. Desor. It was substantially a translation, or rather, a report of verbal information; and it may be considered an accurate sketch of Agassiz's life until his acceptance of the professorship of zoölogy and geology at Harvard University. The article was reprinted in the "Edinburgh New Philosophical Journal," Vol. XLVI., p. 1. Edinburgh, 1848.

1847. — At about the same time, 10 December, 1847, a *Biographic Notice of Professor Agassiz* appeared in New York, in a pamphlet entitled, "Professor Agassiz's Lectures: The Animal Kingdom," issued by Horace Greeley, the editor of the New York "Tribune." The proofs of this short sketch, of only two pages, were corrected by Agassiz and Auguste Mayor.

Many newspapers and magazines reprinted in part these last two biographical sketches.

2. After his Death.

1874. — *Commemorative Notice of Louis Agassiz*, by Theodore Lyman. This paper, by one of his favourite pupils, was published as an academic eulogy, in the "Annual Report of the Council of the American Academy of Arts and Sciences for 1873," Boston, 13 pages. It is a good sketch, and very complete, considering its limitation to a dozen octavo pages.

1874. — *Obituary, Louis John Rudolph Agassiz,* by Benjamin Silliman, Jr. In "American Journal of Science," Third Series, Vol. VII., pp. 77–80, January, 1874. The author gives extracts from two letters of Agassiz, dated October, 1845, and February, 1846, addressed to Benjamin Silliman, Sr.

1874. — *Notice biographique sur Louis Agassiz,* par Alphonse de Candolle, dans son Rapport comme Président de la Société de Physique et d'Histoire naturelle de Genève pour l'année 1873–1874. Lu le 16 Juillet, 1874. "Mém. Soc. Phys. et Hist. Nat.," Vol. XXIII., pp. 470–478. 4to, Genève, 1874. An excellent and original biography, containing two most interesting letters addressed to M. Louis de Coulon of Neuchâtel, written by Agassiz, from Paris, during the months of March and June, 1832.

1874. — *Louis Jean Rodolphe Agassiz,* by the Duke of Argyll, in the Anniversary Address of the President of the Geological Society of London, "The Quarterly Journal," Vol. XXX., May, 1874, pp. xxxvii–xliii. This academic eulogy is less complete than that by Mr. Lyman.

1874. — *Louis Agassiz,* by Dr. F. Steindachner, in "Die Feierliche, Sitzung der Kaiserlichen Akademie der Wissenschaften am 30 Mai, 1874. Wien, pp. 60–82. The author, who accompanied Agassiz in his last journey around South America, and passed three years in constant and most intimate intercourse with him, has here written an excellent "Nekrolog," by far the best published in German. Steindachner, being an excellent ichthyologist, was able to appreciate at its full value the great and, as he says, the unique knowledge of Agassiz, whom he justly calls his master.

1874. — (Sketch of) *Professor Louis Agassiz,* by Richard Bliss, Jr., in "Popular Science Monthly," Vol. IV., pp. 608–618, with portrait. New York, March, 1874.

1875. — *Sketch of Agassiz,* by L. F. de Pourtalès, in "Harvard Book," by F. O. Waille and H. A. Clark, Vol. I., pp. 342–344, folio, with an excellent portrait, taken from the larger photograph by Sonrel, Cambridge, 1875.

This sketch contains exact and little known information in regard to the storage of the first collections made by Agassiz in America, though the account of his European life is short and somewhat inaccurate ; *e.g.* Agassiz was never at school at Orbe.

1877. — *Louis Agassiz, notice biographique,* par Ernest Favre, in " Archives des Sciences de la Bibliothèque Universelle," May and June, 1877, 53 pages ; issued also separately, Genève. An English translation, made by order of Professor Joseph Henry, the secretary of the Smithsonian, was published in the " Annual Report of the Smithsonian Institution for 1878," pp. 236-261, Washington.

The author was too young to have known Agassiz personally, but he makes good use of the knowledge of his father, Alphonse Favre, an old pupil of Agassiz ; and the biography is original and good, containing extracts from letters of Agassiz to Alphonse Favre, and several anecdotes about Agassiz, when in Switzerland. I may add that M. Ernest Favre asked me to write the biography for the " Bibliothèque Universelle." But at the time, I was passing a winter at Algiers, far from all my notes and books ; and I declined, but promised to furnish him some notes, more especially upon the life and works of Agassiz in America. He begged me to do so, as otherwise he would not undertake the work. I therefore sent him notes, which he acknowledged very courteously in a foot-note on the first page of his *Notice.* But, influenced by his acquaintance with M. Desor, and also by a little difficulty which had occurred between Agassiz and his father, he gave Desor much more credit for the fossil echinoderms than he is really entitled to, curiously reversing the facts, by saying that Agassiz was the collaborator of his assistant and secretary. Agassiz began the work on echinoderms many years before he knew Desor, and before Desor came to Neuchâtel. as his secretary, and worked out the greater part of it until 1846 ; and the part taken by Desor is wholly secondary, and far below Agassiz's in excellence. I am obliged to make this statement because M. E. Favre, in quoting me as furnishing numerous facts, seems to indicate that his opinion of the part taken by Desor, in the publication of the " Catalogue des Echinodermes vivants et fossiles," 1846-1848, is more or less in accordance with my information, which is erroneous. As I have previously said, I saw the

manuscript of the Catalogue, from the beginning, at Paris, in 1846;
and it fell to my lot to finish it, after M. Desor left Paris, in
February, 1847, the memoir appearing in January, 1848, under my
editorship.

1879. — *Scientific Worthies.* — *Jean Louis Rodolphe Agassiz*, in
" Nature," Vol. XIX., pp. 573–576. London, April, 1879. This
anonymous paper is mainly a translation of Dr. Steindachner's
necrology of Agassiz, mentioned above. In the main it is exact,
only it is marred by an unusual number of typographical errors, and
several dates are not given with sufficient accuracy.

1879. — *Louis Agassiz*, par Louis Favre, in the " Académie de
Neuchâtel. Programme des cours pour l'année scolaire 1879–1880."
Neuchâtel, 32 pages, 4to, with an excellent portrait. This is by far
the best academic eulogy of Agassiz. The author knew Agassiz personally, and, although not a naturalist, had lived long enough among
naturalists to be able to appreciate justly the great scientific worth of
Agassiz's numerous discoveries. It contains three letters addressed
to M. Louis de Coulon, and several extracts from notes dictated by
Agassiz. After Mrs. E. C. Agassiz's life of her husband, it is the
most important contribution we possess on the life of the great
naturalist. It is to be regretted that it is very little known in
America and England.

1883. — *Memoir of Louis Agassiz*, 1807–1873, by Arnold Guyot.
Read before the National Academy of Sciences, October, 1877, and
April, 1878, in Washington. 49 pages, 8vo. Printed at Princeton,
New Jersey, and not distributed until April, 1883.

The part relating to Agassiz's stay in the Braun family at Carlsruhe, in 1826, during a vacation, when a student at Heidelberg, is
well written and charming; but the author gives too prominent a
place to the Glacial question, considering the other researches of
Agassiz.

The great objection to Guyot's paper is that it seems to be written
not so much in honour of Agassiz, as to urge his own claim to pre-
tended discoveries of the laminated or ribbon structure of the ice.
To all impartial glacialists, the part taken by Guyot in the Glacial

question seems extremely slight, and does not compare, either with
that played by Venetz, de Charpentier, and Agassiz, or even with
that of Rendu, Hugi, Desor, Vogt, Charles Martins, Daniel Dollfus-
Ausset, and James D. Forbes. Guyot not only waited many years
after the death of Agassiz before publishing his claim, but did not
print and distribute his biography of Agassiz until six years after
reading it before the National Academy, and one year after Desor's
death, and after the disappearance of Forbes, his adversary in the
question.

1885. — *Louis Agassiz, his Life and Correspondence*, edited by
Elizabeth Cary Agassiz. Boston and London, 2 vols., 12mo, 1885.
Two Swiss critics well acquainted with Agassiz and his family have
given the key to a just appreciation of the work. M. Auguste
Glardon, in the " Bibliothèque universelle et Revue Suisse," June,
p. 449, says : " The biography of a celebrated man, more especially
when it is due to his widow, always occasions some suspicions. To
demand from conjugal love a true impartiality would be entirely
unreasonable, and we must expect that a monument erected under
such circumstances will always be more or less a mausoleum." The
other critic, M. Charles Berthoud, in the " Journal de Genève,"
14th December, 1886, after expressing his surprise at the complete
silence about the controversy with Karl Schimper, says : " The
book is a eulogy, a brilliant picture, without shadows, of a brilliant
life, — more than a true portrait. We must wait for a complete
biography of the savant." Even as a family eulogy, the work lacks
now and then proper appreciation of some of the great difficulties
under which Agassiz laboured during the greater part of his life,
and of the extraordinary, but always successful, efforts which he
made at several critical moments to overcome them, and to continue
his herculean labours. There are many parts and some most important acts of Agassiz's life which are not touched, or even hinted at,
by Mrs. Agassiz. The obviously eulogistic purpose of the work,
and its unquestionably partial character, diminish its interest. With
these exceptions, the work of Mrs. Agassiz is most important. It
is all that could reasonably be expected from a wife.

1886. — The daughter of Alexander Braun, Mrs. C. Mettenius,

has published a German translation of Mrs. E. C. Agassiz's work, under the title, "*Louis Agassiz, autoriste deutsche Ausgabe von C. Mettenius.* 1 vol., 8vo. Berlin, 1886.

1887. — *Madame Elizabeth C. Agassiz — Louis Agassiz : sa vie, sa correspondance, traduit de l'anglais par Auguste Mayor.* 1 vol., 8vo, 617 pages, with a remarkably good portrait of Agassiz. Neu-châtel, 1887. With the consent of Mrs. Agassiz, M. Mayor has added several notes and extracts from letters not given in the English version, and he has placed at the end the catalogue of Louis Agassiz's publications. With M. Mayor, as well as with Mrs. Agassiz, it was a work of love ; and the French version is a marked improvement on the original English edition, being more complete, and, in consequence more valuable, notwithstanding the suppression of a few letters found in the English edition, and the omission of all the engravings, and the substitution of a portrait of Agassiz, which, by the way, is far superior to the one published by Mrs. Agassiz.

1892. — *Louis Agassiz,* par Philippe Godet, in "Petite Biblio-thèque Helvétique " ; a popular and patriotic publication, containing the biographies of celebrated Swiss. 16 pages, 12mo. Genève, 1892.

1893. — *Louis Agassiz, his Life and Work,* by Charles Frederic Holder. 1 vol., 12mo, 327 pages. New York, 1893. This is a well illustrated volume, giving a sort of *résumé* of Mrs. Agassiz's work, by a person who did not know Agassiz personally. He has gone so far as to quote what he represents as an extract from Agassiz's reply to an offer made by the Emperor Napoleon of a position in France. The correspondence on the subject is well known, having been repeatedly published in newspapers, both in America and in Europe, while I have given in full, in Chapter XVI., pp. 71–72, Agassiz's answer to the offer of M. Rouland, Secretary of Public Institutions, and there is nothing in it resembling Mr. Holder's quotation, " that his family owed nothing to France but exile and poverty ; and that he prized more highly the spontaneous gratitude and gifts of a free people than the patronage of emperors and the formal regard of nobles." It is an apocryphal letter. Agassiz was too courteous and too much a man of the world to write so rudely, in answer to a very flattering and honourable proposal.

At the end of the volume, Mr. Holder gives a "Bibliography of Louis Agassiz," extremely confused, and too often erroneous, repeating papers, and even volumes, and passing over some of Agassiz's most important contributions. *Per contra*, he attributes to Louis Agassiz a book entitled, "A Journey to Switzerland," with which he had absolutely nothing to do.

3. BOOKS AND PAPERS CONTAINING IMPORTANT FACTS OR ORIGINAL VIEWS OF THE VALUE OF AGASSIZ'S LIFE.

1852. — *Trial of the action* of Edward Desor, Plff. *versus* Chas. H. Davis, Deft. before the Circuit Court of the United States, for the district of Massachusetts; for breach of contract to write a book on the geological effects of the Tidal Currents of the Ocean. Tried before His Honor Peleg Sprague. Boston, 1852, 67 pages. In the Appendix, at p. 53, are the "Award between Edward Desor and Louis Agassiz," and several important letters of Admiral Davis, relating to Agassiz.

1857. — *The Fiftieth Birthday of Agassiz*, by Henry Wadsworth Longfellow. Poem. Reprinted in "Louis Agassiz, his Life and Correspondence," edited by E. C. Agassiz, Vol. II., pp. 544–545; and also in the present work.

1863. — *A Claim for Scientific Property*, by Henry James Clark. 3 pages. Cambridge, July, 1863. Distributed widely among savants and libraries. It is a one-sided view of claims in regard to certain portions of the "Contributions to the Natural History of the United States," by Louis Agassiz.

1868. — *Farewell to Agassiz*. Poem by O. W. Holmes, in the "Atlantic Monthly," Vol. XVI., pp. 584–585. Boston, November, 1865. Written on the occasion of his departure for a journey in Brazil.

1873. — *Addresses and Proceedings at the Agassiz Memorial Meeting at San Francisco*, in "Proceedings California Academy of Sciences," Vol. V., pp. 220–243. December, 1873. Reprinted in

"Louis Agassiz, his Life and Work," by C. F. Holder, pp. 219–267. New York, 1893.

1874. — *Tributes to Professor Louis Agassiz*, in "Proc. Boston Soc. Nat. Hist.," Vol. XVI., pp. 210–237. Boston, January, 1874.

1874. — *The Prayer of Agassiz.* Poem by John G. Whittier, in "Tribune Popular Science," p. 46. Boston, 1874. Reprinted in "The Complete Poetical Works" of Whittier, pp. 383–384. Boston, 1881. Written on the occasion of the opening of the Anderson School of Natural History, at Penikese Island, in July, 1873, when Agassiz called upon his pupils to join in silent prayer, asking God's blessing on their work. Reprinted in the present work.

1874. — *Agassiz at Penikese.* With a portrait. An anonymous article, well written and important, in "Tribune Popular Science," pp. 47–64. Boston, 1874.

1874. — *In the Laboratory with Agassiz*, by a former pupil (Samuel H. Scudder), in "Every Saturday," Vol. XVI., pp. 369–370. Boston, April, 1874. Reprinted in "American Poems," Cambridge, and issued separately, as a leaflet, for the Agassiz Fund, by Mr. Barnard.

1874. — *Recollections of Louis Agassiz, a Chapter of Reminiscences*, by Theodore Lyman, in "Atlantic Monthly," Vol. XXXIII., pp. 586–596. Boston, May, 1874.

1874. — *Elegy of Agassiz*, by James Russell Lowell, in the "Atlantic Monthly," pp. 586–587, published in the chapter of reminiscences, with the preceding paper by Theodore Lyman. Boston, May, 1874. "A long poem," written at Florence (Italy), in February, 1874, which, according to Lowell, "is among his best verses." A very friendly tribute by one who "understood and liked Agassiz better as he grew older." (Letters of Lowell to C. E. Norton, in "Letters of James Russell Lowell," edited by Charles Eliot Norton. Vol. II., p. 114. New York, 1894.)

1875. — *Un naturaliste du dix-neuvième siècle — Louis Agassiz*, par Émile Blanchard, in "Revue des Deux mondes," July and August, 1875, 64 pages. Paris. Issued also separately as a pamphlet.

This is a scientific sketch rather than a biographic notice. The author, an entomologist and a writer of scientific articles in the " Revue des Deux mondes," although slightly acquainted with Agassiz, knew little of his life and of his great works on zoölogy, palæontology, and the glaciers. The article is sympathetic, and written for a special class of readers, not savants, but *dilettanti;* but there is nothing new or impressive in it. M. Blanchard takes this occasion to claim *priority* over Agassiz in regard to the differences presented between species on account of a more or less advanced state of development and the diminution of typical characters among small species of great natural families.

1878. — *Jean de Charpentier,* par le Docteur Herman Lebert, a biography published in " Actes de la Société Helvétique des Sciences naturelles réunies à Bex," août, 1877, pp. 140–154. Lausanne, 1878. In it, two pages, pp. 149–150, are devoted to Louis Agassiz; remarkably correct and of a fine touch; an excellent sketch, by one who had known Agassiz intimately, and was the first savant who associated with de Charpentier in the glacial theory. His reminiscence of Agassiz, with its interesting anecdotes, is most important, and one of the finest tributes to both de Charpentier and Agassiz.

1879-1887. — *Recollections of Agassiz,* by Edwin P. Whipple, in his " Character and Characteristic Men," and " Recollections of Eminent Men," pp. 266–292, and pp. 77–118. 8vo; Boston. 1879 and 1887. An excellent critique, the best from a literary point of view.

1881. — *Louis Agassiz, son activité à Neuchâtel comme naturaliste et comme professeur de 1832 à 1846,* par Louis Favre. " Bulletin Soc. Sc. nat. de Neuchâtel," Vol. XII., pp. 355–372. Neuchâtel, 2 Juin, 1881. " Un hommage tardif," as it is called by its author. At the time of Agassiz's death, the president, M. Louis de Coulon, announced the painful news, at the meeting of the 18th December, 1873, simply saying, " M. Agassiz jouissait au milieu de nous de l'estime générale." That was all. And the society founded by Agassiz, in 1832, waited eight years before a eulogy of him was read before it, a " devoir sacré," as it was called by Louis

Favre. The article is good and highly complimentary, but rather " tardif."

1882. — *Eduard Desor. Lebensbild eines Naturforscher,* von Carl Vogt. Breslau. 37 pages. 8vo. Published first in a review, it was afterwards issued as part of the " Deutsche Bücherei." Several pages of this biography of Eduard Desor are filled with notes on Agassiz. Although rather prejudiced, and written in a tone of severe criticism, and entirely hostile to the inhabitants of Neuchâtel, the article contains interesting and generally fair accounts of the life of Agassiz at Neuchâtel from 1839 to 1844, and on the glacier of the Aar from 1840 to 1843. This was during the most active part of Agassiz's scientific life, and at a very critical period in the publication of his costly works. Vogt, with his sharp eyes, inclined to see the humorous of everything, gives a rather piquant inside view of Agassiz's scientific and business methods. The article is written somewhat coarsely, but humorous, and not always in sympathy with his subject ; for even his old companion Desor is not secure from his scorching criticism.

1882. — *Alexander Braun's Leben nach seinem handschriftlichen Nachlasz,* dargestellt von C. Mettenius. Berlin, 1882. This most interesting biography of Agassiz's friend and brother-in-law, Alexander Braun, contains several letters by or to Agassiz, and many references to their relations when students at Heidelberg, Munich, and Paris. Although written by his daughter, Alexander Braun's life is not a family eulogy, but a true life, with many private incidents, which give a tone of veracity, most appreciated by those who want to know in full the different phases of the character of a savant.

1882. — *Histoire abrégée de la Société Neuchâteloise des Sciences Naturelles depuis sa Fondation,* par Louis Favre. " Bull. Soc. Sc. Nat. de Neuchâtel," Vol. XIII., pp. 3–33. Neuchâtel, Décembre, 1882. This is an address on the fiftieth anniversary of the foundation of the Natural History Society of Neuchâtel, — " société dont l'idée et l'initiative sont dues à Agassiz," according to the author. It contains important facts in the scientific life of Agassiz during his stay at Neuchâtel. It was reprinted under the title, "Cinquante-naire de la Société Neuchâteloise des Sciences naturelles," in "Musée Neuchâtelois," Vol. XX., pp. 84–90, and pp. 99–112. 4to. February and March. Neuchâtel, 1883.

1883.— *Louis Agassiz at Neuchâtel,* by Jules Marcou. "The Nation," Vol. XXXVI., pp. 36, 4to. Jan. 11, New York, 1883. We have in this an exact and most complete list of countries where ancient glaciers have been found ; proving the existence of a "Glacial epoch," as prophesied by Agassiz at Neuchâtel, in July, 1837.

1886.— *Louis Agassiz. Étude biographique,* par Auguste Glar-don, in "Bibliothèque universelle et Revue Suisse." Third series, Vol. XXX., June and July, 1886, pp. 449–481, and 116–146. Lau-sanne, 1886.

The author, a Vaudois, has known the Agassiz family for the last three generations, and his critical review gives a good and true account of Louis Agassiz. I will quote his own impression of Agassiz's departure from Neuchâtel for America : " Il était deux heures du matin, lorsque le professeur quitta la maison qui avait pendant treize ans abrité son bonheur domestique et ses collections. Les étudiants vinrent en corps lui donner une sérénade d'adieu à la lueur des flambeaux ; ses collègues de l'Académie étaient aussi présents. L'émotion était générale ; plusieurs avaient le pressenti-ment que l'Amérique retiendrait le professeur aimé et qu'on ne le reverrait jamais."

1886.— *Glaciers and Glacialists,* by Jules Marcou. "Science," Vol. VIII., pp. 76–80, 4to. July 23, New York, 1886. This is an explanation of the glacial doctrine, with dates of the discoveries, and an account of the part taken by Louis Agassiz.

1886.— Of all the numerous articles in newspapers, American or foreign, reviewing the work of Mrs. Agassiz in the English, German, and French versions, I shall quote only one containing original suggestions and facts not recorded in the work. It is a review in the "Journal de Genève" of the 14th December, 1886, entitled : *Louis Agassiz, sa vie et sa correspondance,* by Charles Berthoud, an old friend of Agassiz, who freely speaks his impressions of the book and its contents in the words, " Ce livre n'est point une biographie scientifique," and he regrets to find so few letters of Agassiz on scientific subjects.

1887. — *Das 50 jährige Jubiläum der Eiszeit-Lehre,* 1837, 15 Feb. 1887, von Dr. Otto Volger, in " Beilage zur Allgemeinen Zeitung." München, pp. 697, 698, and pp. 715, 716. Folio, Munich, 1887. .The facts, although quite correct in the main, give only the views on one side, in favour of Karl Schimper's claim ; and the article is unjust in regard to Agassiz's part in the controversy.

1887. — *Il Naturalista Agassiz secondo le memorie scritte da sua Moglie,* by Paolo Lioy, in " Nuova Antologia." Terza serie, Vol. VIII., pp. 240–258. Roma, Italy, 16 Mazo, 1887. A remarkable and fascinating article, full of life, and written with such a *brio* that Agassiz seems to pass before the eyes.

1887. — *Souvenir de l'inauguration du buste élevé à L. Agassiz par la Société des Belles-Lettres dans le Bâtiment Académique de Neuchâtel, le* 12 *Mai,* 1887. 65 pages, 8vo, with portrait. Neuchâtel, 1887. Several of the speeches and addresses delivered during the ceremony contain new facts and special characteristics, as well as anecdotes in regard to the great naturalist.

1889. — *La première Académie de Neuchâtel. Souvenirs de* 1838– 1848, par Alphonse Petitpierre. 12mo. Neuchâtel, 1889. This contains several extremely interesting letters of Agassiz. The author shows the great part taken by Agassiz in the founding of the Academy, and its prosperity so long as he inhabited Neuchâtel.

1892. — *Agassiz at Penikese,* by David S. Jordan, in " Popular Science Monthly," Vol. XL., pp. 721–729. New York, April, 1892. An interesting reminiscence of the school at Penikese Island in 1873, by one of the pupils.

Post-scriptum. — Finally, Jean Jacques Antoine Ampère in his book, *Promenades en Amérique* (2 vols., 1852), speaks of his stay at Agassiz's house in Cambridge ; and Auguste Laugel, some years after (1865), in a volume on *Les États-Unis pendant la guerre de*

Sécession, gives an account of a visit to Agassiz. The last author has also published in the " Revue de Deux mondes," of 1857, an article entitled : *M. Agassiz et ses travaux*, Vol. XI.. pp. 77–108. The celebrated Julius Froebel, of kindergarten fame, in his work *Aus Amerika*, 1857, relates a visit to Agassiz.

Second *Post-scriptum.* — 1863. — *Étude sur l'industrie huitrière des États-Unis*, par Philippe de Broca, contains a letter of Agassiz and notes on the acclimatization of oysters, pp. 4–6.   Extrait de la " Revue Maritime et Coloniale," Paris, 1863.

1869. — *De la Science en France*, par Jules Marcou, contains an interesting letter of Agassiz, and a list of Agassiz's principal publications, pp. 185–191, Paris, 1869.

APPENDIX B.

───◆◆───

AGASSIZ'S PORTRAITS, ENGRAVINGS, PHOTOGRAPHS, BUSTS, MEDALS, AND TABLETS.

AGASSIZ'S face has been popularized by many portraits, although not a single good oil portrait of him exists. The only good coloured picture is a miniature pastel drawing, made by his first wife, when he was a student at Heidelberg, at the age of nineteen. A copy of it forms the frontispiece of Vol. I., of " Louis Agassiz, his Life and Correspondence," edited by Elizabeth Cary Agassiz.

In 1840, Jacques Burkhardt made a portrait of Agassiz, but it is a very poor likeness. It is preserved in the public library of Neuchâtel.

In 1842, Agassiz's portrait was painted, of natural size, by an artist of the name of Zuberbühler. His face is much coloured, as if by sunburn, and he is dressed in a brown coat, decorated with a long golden chain, and is represented surrounded by masses of ice. It is a poor likeness, and the picture as a whole is not in good taste. The original is now at the house of Mr. Alexander Agassiz at Cambridge. A copy of it, made at Neuchâtel for M. Auguste Mayor, has been improved in regard to the surroundings; instead of a great mass of ice with blue-green crevices, the background is occupied by a true landscape of the glacier of the Aar, and showing the Finsteraarhorn, the Agassiz's horn, and the " Hôtel des Neuchâtelois."

In 1846, M. Fritz Berthoud, a banker of Neuchâtel, at the same time an amateur painter, then a resident of Paris, made a full-length

picture of Agassiz and Desor on the same canvas; neither is a
good likeness, that of Agassiz more especially being very poor.
This large picture is now in the fine picture gallery of the city of
Neuchâtel.

In 1886, another oil portrait, by Alfred Berthoud, by order of
the Canton of Neuchâtel, was painted, and placed, first, in the hall
of the Great Council of the canton, and afterward in the *Aula* of
the Academy of Neuchâtel.

After Agassiz's death, in 1875, a large oil portrait was made by
Mrs. C. V. Hamilton, and is placed in the library of the Boston
Society of Natural History. The likeness is not good, and it is a
very poor representation of the great naturalist. A painter named
Billings, also, made an unsuccessful attempt at a picture of Agassiz.

Lately, 1894, another three-quarters length life-sized portrait of
Agassiz has been executed in oil, by an American artist, Walter
Gilman Page, who never saw him when he was alive. The flesh
tints are far too exaggerated, and the picture does not give a correct idea of the original. It also is an unsatisfactory likeness. It
has been placed at the Agassiz School in Jamaica Plain, near Boston.

If we do not possess a single good likeness in oil of Agassiz, we
have, *per contra*, many excellent lithographs and photographs.
The first one is to be found in "Excursions et séjours dans les
glaciers et les hautes régions des Alpes, de M. Agassiz et de ses
compagnons de voyage," par E. Desor, Neuchâtel, 1844. It is the
frontispiece of the volume, and was drawn on stone, by A. Sonrel,
from a daguerrotype of very small size. The likeness was not
very good, except the upper part of the head.

The second was published in the "Album de la Suisse Romane,"
Geneva, to accompany his biography, page 1, by his friend Jules
Pictet de la Rive, in Vol. V., 1847. The drawing was made on
stone, from life, by M. P. Élie Bovet, at Neuchâtel, in 1845. It is
a good portrait, cabinet size; rather rare. It was unknown to his
family, as well as the biography, until I discovered them in 1887,
while reading Agassiz's correspondence with Pictet.

The third portrait appeared as the frontispiece of the first volume
of "The Annual of Scientific Discovery; or, Year-Book of Facts
in Science and Arts," edited by David A. Wells and George Bliss,

1849, Boston. It was drawn on stone, by A. Sonrel, from a daguerrotype. The likeness is not satisfactory. This is the first time
that a facsimile of his signature was published under the portrait.

After 1859, many photographs were taken, more especially by
A. Sonrel. All are good likenesses. I shall mention only the
larger ones. A large sized one was taken in 1863, and has circu-
lated much among his friends and students. A reduction of it is
engraved as a frontispiece of Agassiz's "Geological Sketches,"
Boston, 1870.

Another full-length photograph was taken in 1869, representing
Agassiz, Professor Benjamin Pierce, then Superintendent of the
United States Coast Survey, and Captain Carlisle P. Patterson,
Chief Hydrographer of the Coast Survey; all three are seated.

The same year, 1869, another large photograph represented
Agassiz seated and looking at a globe, on which Professor Pierce,
who is standing, points out the Gulf Stream above the Pourtalès
Plateau. Both portraits are excellent. The one of Agassiz, representing him almost in profile, has been reproduced since by Justin
Winsor in his Vol. I., p. 373, of his "Narrative and Critical History of America."

A cabinet photograph was taken at San Francisco by Watkin,
in 1872, just after his arrival from his voyage in the *Hassler*. A
good photograph of Mrs. Agassiz was made at the same time.

Another cabinet photograph, in 1872, the last made by Sonrel,
is a splendid profile of Agassiz; it was taken especially for the
engraving of the large bronze medal at Neuchâtel, by Professor F.
Landry. I give it as the frontispiece of Vol. I.

Among the numerous *cartes de visite*, I may mention one taken
in 1863, representing Agassiz seated, with manuscript in his left
hand; it is remarkably well executed, showing his peculiar attractive smile and brilliant eyes. Another taken at the same time,
represents him in front of the blackboard and lecturing before
his pupils, with an echinide drawn in white chalk on the blackboard. Eight years later another photograph represented Agassiz
and Pourtalès together seated at a table, on which lie a book, a
stone, and specimens of echinoderms; Agassiz holds an echinus
in his left hand, and in the right hand a lens, through which he is

looking attentively at the specimen. This portrait, with a few changes, has been used by Mr. C. F. Holder, as the frontispiece for his volume, " Louis Agassiz, his Life and Work," New York, 1893 : but the engraving is very poor, and the likeness decidedly bad. In 1866, when at Rio de Janeiro, just after his return from the Amazons, Agassiz was taken at full length with his friend Major Coutinho, Agassiz's right hand resting on the right shoulder of Coutinho. This is one of the most animated portraits of Agassiz, who looks browned by his ten months' stay on the great Amazons, but full of life and very spirited, with his piercing eyes and his strong frame, so much in contrast with Coutinho's small size.

The portrait of Agassiz, forming the frontispiece of Vol. II. of Mrs. Agassiz's life of her husband, is taken from an engraving, which appeared in "Nature," April, 1879, with a biography of Agassiz, one of the " Scientific Worthies Series " of that periodical. The likeness is poor. But in the French translation of Mrs. Agassiz's work, by Auguste Mayor, the frontispiece portrait of Agassiz is excellent, the best by far of all those published. The portrait published by Louis Favre, in his biography of Agassiz, forming part of the " Programme des cours de l'Académie de Neuchâtel pour l'année scolaire. 1879, 1880," is also a good likeness.

I know only one double photograph of Agassiz for use in a stereoscope. It was made by Sonrel in 1861, and represents Agassiz in his library at his home in Quincy Street. A part of the library is visible, as well as a geological map of Central Europe hanging against the bookcase. It was made as an imitation of " Alexander von Humboldt in his library," a popular engraving often seen in Germany and Switzerland, and a part of which may be seen in the corner of the photograph. Agassiz is seated at his desk. loaded with manuscripts, and looking through a magnifying-glass at a fossil on a small stone held in his left hand. The expression is rather too serious, but it is a good portrait.

As to the three' busts executed after his death, by three artists who had never seen him. they are all poor so far as likeness is concerned. One is by Mr. Preston Power, and may be seen at the Agassiz Museum, and a cast of it at the library of the Boston

Society of Natural History; the second, by Henry Dexter, is in the gallery of the Museum of the Boston Society of Natural History; the third, by M. Iguel of Neuchâtel, was erected with appropriate ceremonies, May 12, 1887, at the Academy of Neuchâtel, by the " Société de Belles-Lettres " of Switzerland, which raised a subscription among its members to cover the expenses. This bust, by M. Iguel, although well executed and more elaborate than the others, does not give a true likeness of Agassiz, not even so good as the one by Power. It is placed on a pillar of brown marble, on which is engraved : " A Agassiz, la Société de Belles-Lettres, 1887."

MEDALS AND TABLETS.

The medals executed, one at Neuchâtel and the other at the National Mint in Philadelphia, are both good. The one engraved at Neuchâtel and coined at Geneva in 1876 is very remarkable, both on account of its execution and its size; it is one of the best medals ever struck, being so large as to look like a medallion, and is most creditable to the engraver, Professor Fritz Landry, of Neuchâtel. The module or size is 94 mill. On the obverse, the legend is Ls. Agassiz, 1807-1873, — F. Landry, Neuchâtel, Suisse. On the reverse we read as exergue enclosed in a crown of laurels : *Viro ingenio labore scientia Præstantissimo.* It is a bronze medal, of which one hundred and fifty-one copies were struck, and two copies in silver by special request.

The other, engraved in 1875 by W. Barber, an artist at the Philadelphia Mint, is much smaller. The size is 45 mill., and is the one used by the National Government for all medals struck to honor the memory of great men in America. The medal taken as a model for that series is the Benjamin Franklin medal, engraved by A. Dupré, in 1784, at Paris. The size is rather small, which gives to all these medals an unattractive appearance. The profile of Agassiz is good, but the details are not so harmonious and exact as they were in nature and in the photograph used by the engraver. On the obverse, the legend is simply *Agassiz*, without any of his Christian names. On the reverse, we read as exergue *na.* 1807, *ob.* 1873, and as legend : *Terra Marique Ductor indagatione naturæ.* This medal also is bronze. During 1876 and 1877 only thirty-one copies were struck; and, in 1879, a silver one was struck, according to the reports of the director of the Mint.

Tablets to the memory of Louis Agassiz have been placed in Europe and in America. The inhabitants of his birthplace placed over the door of the parsonage of Motier (Fribourg) a marble tablet with the inscription: " J. Louis Agassiz, célèbre naturaliste est né dans cette maison, le 28 Mai, 1807 "; and the Cornell University at Ithaca, New York, unveiled a marble tablet, in commencement week, June, 1885, in the founder's chapel. At the opening of the University in 1868, Agassiz was present, made a speech, and immediately after began a course of twenty lectures before a very large audience, including almost all the professors, instructors, and students. The inscription on the tablet reads in black lines as follows: " To the Memory of Louis Agassiz, 1807–1873. In the midst of great labors for science throughout the world, he aided in laying the foundation of instruction at the *Cornell University*, and by his teaching here gave an impulse to scientific studies which remains a precious heritage. The Trustees, in gratitude for his counsels and teachings, erect this memorial, 1884."

In September, 1885, a large stone slab was placed by his son in the wall of the entrance hall of the Museum of Comparative Zoölogy at Harvard University, with the following inscription: " Ludovici — Agassiz — Patri — filius — Alexander — MD — CCC — LXXX."

An Agassiz Memorial Fund was subscribed during 1874 and 1875, to be used for the completion of his Museum. $245,792 were received, of which $130,000 came from his son and his daughter Pauline, and $9192 from teachers and pupils, while the state granted $50,000, the total amounting to $310,600.

But the most original memorial is the inscription on the boulder of micaceous schist, once forming part of the " Hôtel des Neuchâtelois," on which is engraved in large letters *L. Agassiz* above the name of *Hôtel des Neuchâtelois*, 1840.

In 1844, the roof and sides of the " Hôtel des Neuchâtelois " broke apart, and afterward frost divided the boulder into a thousand pieces. Happily, Edouard Collomb, in 1842, had drawn, in water colour, the north face of the block on which were engraved the names of all the assistants. Daniel Dollfus-Ausset has since published the picture in his Atlas of the " Matériaux pour l'étude des glaciers "; and, at my suggestion, a small, but good, reproduction of it was given in "Science," Vol. IV., p. 360, October, 1884, Cambridge. A correct and successfully executed reproduction of Collomb's water colour picture of the "Hôtel des Neuchâtelois " is given in Vol. I., opposite p. 202. In August, 1884, several pieces with inscriptions on them were found at a great distance : twenty-four hundred metres lower than the position of the hotel, as determined by Agassiz in 1842, giving an average annual velocity of fifty-five meters. Thus, many years after his death, Agassiz, through his inscriptions on the boulder, is still the promoter of valuable discoveries on the Aar glacier.

APPENDIX C.

---◆◇◆---

1828. — 1. *Cynocephalus Wagleri.* — *Isis*, 1828, Part IX., pp. 861–863, with a Fig. Translated into French and reprinted in *Férussac, Bull.*, Vol. XIX., 1829, pp. 345–346, under the title "Description d'une nouvelle espèce du genre Cynocéphale."

1828. — 2. Beschreibung einer neuen species aus dem genus *Cyprinus*. (*Cyprinus uranoscopus*) nouvelle espèce trouvée par Agassiz à Munich, et présentée à la réunion des savants d'Allemagne à Berlin par Oken. — *Isis*, 1828, Part X., pp. 1046–1049, and *Isis*, 1829, Parts III. and IV., pp. 414–415. French translation in *Férussac, Bull.*, Vol. XIX., 1829, pp. 117–118.

1829. — 3. Selecta genera et species piscium quas in itinere per Brasiliam annis 1817–1820 collegit et pingendos curavit J. B. de Spix; digessis, descripsit et observationibus anatomicis illustravit Dr. L. Agassiz ; præfatus est et edidit itineris socius Dr. de Martius. Monachii, 1829, folio with 29 plates. Reviewed in *Isis*, 1829, Part VII., p. 715.

1830. — 4. Dissertatio inauguralis : De taxi et syntaxi morphomatum telæ corneæ dictæ. 4to, Monachii, 1830. This title is in the centre of the second page. On the first page used as a cover

---

[1] Several short papers appear more than once in this list, because they are either extracts from some of the larger works of Agassiz, or because he was in the habit of sending the same paper, with only a few words changed, to several scientific periodicals. This is unavoidable in a complete Bibliography.

we read: Ad Disputationem publicam [under the presidency of Roeschlaub]. Pro summis in medicina chirurgia et arte obstetricia honoribus rite obtinendis a prænobili, clarissimo et doctissimo viro ac domino Ludovico Agassiz, A. A. LL. Philos. Doct. Urbigenensi, Helveto, Die III, Aprilis, MDCCCXXX. habendam, etc. There are 74 theses. Thesis I, page 3, entitled: Fœmina humana superior mare, caused a sensation among the examiners and the audience, for Agassiz proved that the organization of woman was more complicated and superior to that of man.

1830. — 5. Prospectus de l'"Histoire naturelle des Poissons d'eau douce de l'Europe centrale, ou Description anatomique et historique des Poissons qui habitent les lacs et les fleuves de la chaine des Alpes et les rivières qu'ils reçoivent dans leurs cours." Small folio, Munich, 1830. This rare prospectus preceded by twelve years the publication of the first part of the work, which will be found in its chronological place, No. 90, 1842. An announcement of the proposed work is inserted in *Férussac, Bull.*, Vol. XXIII., p. 271. 1830.

1832. — 6. Untersuchungen über die fossilen Süsswasser-Fische der tertiären Formation. *Leonhard und Bronn, Jahrb.*, pp. 129–138. 1832.

1832. — 7. Untersuchungen über die fossilen Fische der Lias formation. *Leonhard und Bronn, Jahrb.*, pp. 139–149. 1832.

1833. — 8. Tableau synoptique des principales familles des plantes. 12mo. Neuchâtel, 1833.

1833–1844. — 9. Recherches sur les Poissons fossiles, 5 Vols. 4to, with four hundred coloured folio plates. Neuchâtel, 1833–1844. The work was noticed in *Isis*, 1834, Part IV., p. 405, and 1835, Part II., p. 135; in *Leonhard und Bronn, Jahrb.*, 1834, pp. 242, 484; 1835, p. 595, etc.; 1844, p. 250. Also in *Silliman's Amer. Journ. Sc.*, Vol. XXVIII., p. 193, and Vol. XXX., p. 34.

1833–1835. — 10. Résumé des travaux de la section d'histoire naturelle, et de celle des sciences médicales (de la Société des sciences naturelles de Neuchâtel) pendant l'année 1833. *Mém. Soc. sc. nat. Neuchâtel*, Vol. I., pp. 17–28, Neuchâtel, 1835. Although the volume was not distributed until 1835, the report was issued among the resident fellows at the end of 1833.

1833. — **11.** Neue Entdeckungen über fossile Fisches. *Leonhard und Bronn, Jahrb.*, pp. 675–676. 1833. Also in *Edinb. New Phil. Journ.*, Vol. XXXVII., p. 331.

1833. — **12.** Synoptische Uebersicht der fossilen Ganoiden. *Leonhard und Bronn, Neues Jahrb.*, 1833, p. 470.

1834. — **13.** Remarks on the different species of the genus *Salmo* which frequent the various rivers and lakes of Europe. *Report, British Assoc. Adv. Sc. Edinburgh*, pp. 617–623. 1834. *Edinb. New Phil. Journ.*, Vol. XVII., pp. 380–385. 1834. *L'Institut*, Vol. III., pp. 72-73. Paris.

1834. — **14.** On the fossil fishes of Scotland. *Report British Assoc. Adv. Sc. Edinburgh*, pp. 646–649. Reprinted in *L'Institut*, Vol. III., No. 94, pp. 65–66. Paris, 1835.

1834. — **15.** On a new classification of fishes, and on the geological distribution of fossil fishes. *Proc. Geol. Soc. London*, Vol. II., No. 37, pp. 99–102. November 5, 1834. Also in *London and Edinb. Phil. Mag.*, Vol. V., pp. 459–461. London, 1834. And in *Edinb. New Phil. Journ.*, Vol. XVIII., pp. 175–176. 1835.

1834. — **16.** On the anatomy of the genus *Lepidosteus*. *Proc. Zoöl. Soc. London*, Vol. IV., pp. 119–120. 1834. Also in *L'Institut*, Vol. III., p. 190. Paris, 1835.

1834. — **17.** Observations on the growth and the bilateral symmetry of the Echinodermata. *London and Edinb. Phil. Mag.*, New Series, Vol. X., pp. 369–373. London, 1834. Reprinted under the title: Ueber die äussere Organisation der Echinodermen in Oken, *Isis*, pp. 254–257. 1834.

1834. — **18.** Ueber das Alter der Glarner Schiefer-Formation nach ihren Fischresten. *Leonhard und Bronn, Neues Jahrb.*, 1834, pp. 301–306.

1834. — **19.** Allgemeine Bemerkungen über fossile Fische. *Leonhard und Bronn, Neues Jahrb.*, 1834, pp. 379–390.

1834. — **20.** Résumé des travaux de la section d'histoire naturelle, et de celle des sciences médicales de la Société des sciences naturelles de Neuchâtel, pendant l'année 1834. *Mém. Soc. sc. nat. Neuchâtel*, Vol. I., pp. 28–32. 1834.

1834. — 21. Description de quelques espèces de Cyprins du lac de Neuchâtel, qui sont encore inconnues aux naturalistes. *Mém. Soc. sc. nat. Neuchâtel*, Vol. I., pp. 33–48. Neuchâtel, mai, 1834. *L'Institut*, Vol. IV., pp. 419–420. Paris, 1836.

1835. — 22. Notice sur les fossiles du terrain crétacé du Jura Neuchâtelois. *Mém. Soc. sc. nat. Neuchâtel*, Vol. I., pp. 126–145. 1835. *L'Institut*, Vol. IV., pp. 420–421. Paris.

1835. — 23. Prodrome d'une monographie des Radiaires ou Echinodermes. *Mém. Soc. sc. nat. Neuchâtel*, Vol. I., pp. 168–199. 1835. (Read the 10 January, 1834.) *Ann. sc. nat. Zoologie*, pp. 257–296. Paris. *Ann. nat. Hist.* or *Mag. Zoöl. Bot. & Geol.*, Vol. I., pp. 30–43, 297–307, 440–449. London, 1838.

1835 — 24. Sur les Belemnites (mémoire communiqué à l'académie des Sciences par Férussac). *Comptes Rendus Acad. sc. Institut de France*, Vol. I., p. 341. Paris. Reprinted under the title, "Ueber Belemniten." *Leonhard und Bronn, Neues Jahrb.*, 1835, p. 168.

1835. — 25. Revue critique des Poissons fossiles figurés dans l'*Ittiolihologia Veronese*, Neuchâtel, 1835. *Leonhard und Bronn, Neues Jahrb.*, 1835, pp. 290–316.

1835. — 26. Sur les poissons fossiles de la formation houillère. *L'Institut*, Vol. III., pp. 253–254. Paris, 1835.

1835. — 27. On the principles of classification in the animal kingdom in general, and among mammalia in particular. *Report British Assoc. Adv. Sc.*, *Dublin*, pp. 67–68. 1835.

1835. — 28. Systematic enumeration of the fossil fishes in English collections. *Proc. Geol. Soc. London*, Vol. II., pp. 207–208. Nov. 7, 1835. Translated into French, under the title, "Sur les Poissons fossiles de l'Angleterre," *L'Institut*, Vol. IV., pp. 85–86. Paris, 1836.

1835. — 29. Remarques sur les Poissons fossiles. *Bull. Soc. natur. Moscou*, Vol. VIII., pp. 180–201. Moscou, 1835.

1835. — 30. Coup d'œil synoptique des Ganoïdes fossiles. *Bull. Soc. natur. Moscou*, Vol. VIII., pp. 202–318. · Moscou, 1835.

1835. — 31. Views of the affinities and the distribution of the Cyprinidæ. *Proc. Zoöl. Soc.*, 1835, pp. 149–151. London.

1835. — 32. On the arrangement and geology of fishes. *Edinb. New Phil. Journ.*, Vol. XIX., pp. 331–346. 1835.

1835. — 33. Observations sur les blocs erratiques des pentes du Jura. *Bull. Soc. géol. France*, Vol. VII., p. 30. Paris, novembre, 1835.

1835. — 34. Notice on the fossil beaks of four extinct species of Fishes, referrible to the genus *Chimæra*, which occur in the oolitic and cretaceous formations of England, by W. Buckland, with the distinctive characters of each species, by L. Agassiz. *Proc. Geol. Soc. London*, Vol. II., pp. 205–206. London, November, 1835. *London and Edinb. Phil. Mag.*, Vol. VIII., pp. 4–7. 1836.

1836. — 35. Résumé des travaux de la Société des Sciences naturelles de Neuchâtel, Section d'histoire naturelle et de médecine, de 1834 à 1836. *Mém. Soc. sc. nat. Neuchâtel*, Vol. II., pp. 6–11. Distributed to the fellows in 1836, and issued with the volume in 1839. Neuchâtel.

1836. — 36. Les Poissons fossiles de l'Angleterre. *L'Institut*, Vol. IV., pp. 85–86. Paris, 1836.

1837. — 37. Discours prononcé à l'ouverture des séances de la Société Helvétique des sciences naturelles, à Neuchâtel le 24 juillet, 1837, par L. Agassiz, Président. *Actes de la Soc. Helvétique sc. naturelles, Neuchâtel, 24, 25 et 26 juillet*, 1837, 22ᵉ session, pp. v–xxxii. Neuchâtel, 1837.

1837. — 38. Des glaciers, des moraines et des blocs erratiques. *Bibl. univ. Genève*, Vol. XII., pp. 369–394. *Edinb. New Phil. Journ.*, Vol. XXIV., pp. 364–383. Edinburgh.

1837. — 39. Sur les blocs erratiques du Jura. *Comptes Rendus Acad. sc. Paris*, Vol. V., pp. 506–508. Also in *Edinb. New Phil. Journ.*, Vol. XXIV., p. 176–179. *Bibl. univ. Genève*, Vol. XI., pp. 416–418. *L'Institut*, Vol. XI., pp. 417–418. Paris.

1837.—**40.** Sur les infusoires fossiles du tripoli d'Oran. *L'Institut*, Vol. V., pp. 330–331. Paris, 1837.

1837.—**41.** A systematic and stratigraphical catalogue of the fossil fish in the cabinets of Lord Cole and Sir Philip Grey Egerton, by Sir Philip Grey Egerton. With notes of his system of Ichthyology, hitherto uncommunicated to the public, by L. Agassiz. The synoptical table of the orders and families is in French, by Agassiz. 23 pages. 4to. London, 1837.

1837–1844. — **42.** Mineral-Conchologie Grossbrittaniens, von James Sowerby; deutsche Bearbeitung, herausgegben von Hercules Nicolet, durchgesehen von Dr. Agassiz. 1 livraison. The work was not finished until 1844, when it appeared under the title, "James Sowerby's Mineral-Conchologie Grossbrittaniens oder ausgemalte Abbildungen und Beschreibungen der Schalthier-Ueberreste welche zu verschiedenen Zeiten und in verschiedenen Tiefen der Erde erhalten worden sind. Deutsch bearbeitet von Ed. Desor. Durchgeschen und mit Anmerkungen und Berichtigungen versehen von Dr. Louis Agassiz. 395 colorite Tafeln und 88 halbe Bogen Texte. Neuchâtel und Solothurn, 1837–1844.

1837–45. — **43.** Conchyliogie Minérale de la Grande Bretagne, par James Sowerby, traduit de l'anglais par E. Desor, avec un avant-propos et des notes par L. Agassiz. 1 vol. text, 1 vol. coloured plates. Neuchâtel and Soleure, 1837–45.

1838. — **44.** Notice sur les moules du Musée de Neuchâtel. 3 p. 4to. Neuchâtel, novembre, 1838.

1838.—**45.** Künstliche Steinkerne von Konchylien-Fisches. *Leonhard und Bronn, Neues Jahrb.*, pp. 49–51. 1838.

1838. — **46.** Theorie der erratischen Blöcke in der Alpen. *Leonhard und Bronn, Neues Jahrb.*, pp. 303–304. 1838.

1838. — **47.** Monographies d'Echinodermes vivans et fossiles. I^bre livraison, *Les Salénies*. 4to. Neuchâtel, 1838.

1838. — **48.** Discussions sur les argiles de Speeton (Yorkshire); le *Spatangus retusus*; les causes de modification des êtres vivants;

et les ossements fossiles de Stonesfield. *Bull. Soc. géol. France,*
Vol. IX., pp. 262–266.   Paris, avril, 1838.

1838. — **49.** Observations sur les glaciers. *Bull. Soc. géol. France,*
Vol. IX., p. 407.   The paper was published farther on, at pp. 443–
450, under the title, " Note de M. Agassiz sur les glaciers." Reproduced in the *Bibl. univ. Genève,* Vol. XX., p. 382 (1839) ; and also
in *Excursions dans les Alpes,* par E. Desor, pp. 1–14, under the title,
" Notice sur les glaciers " (1844).   Translated into English, under
the title, " Remarks on Glaciers, Read at a Meeting of the Geological
Society of France," and published in *Edinb. New Philos. Journ.,*
Vol. XXVII., pp. 383–390.   Edinburgh, October, 1839.

1838. — **50.** Réponses aux objections du transport des blocs
erratiques par la marche des glaciers, et du poli des roches. *Bull.
Soc. géol. France,* Vol. IX., p. 409.   Porrentruy, septembre, 1838.

1838. — **51.** Conjectures sur l'origine des couches spatiques observées dans le Jura. *Bull. Soc. géol. France,* Vol. IX., p. 426.
Porrentruy, septembre, 1838.

1838. — **52.** Explications sur les laves vues à la Neuveville.
Bull. Soc. géol. France, Vol. IX., p. 435.   La Neuveville, septembre,
1838.

1838. — **53.** Le terrain néocomien plus récent que la formation
wealdienne. *Bull. Soc. géol. France,* Vol. IX., p. 435.   Ile St Pierre,
septembre, 1838.

1839. — **54.** Notice sur quelques points de *l'organisation des
Euryales,* accompagnée de la description détaillée de l'espèce de la
Méditerranée, 14 pages, with 5 plates. *Mém. Soc. sc. nat. Neuchâtel,*
Vol. II.   Neuchâtel, 1839.

1839. — **55.** Notice sur le *Mya alba,* espèce nouvelle de Porto-
Rico.   2 pages, with a plate. *Mém. Soc. sc. nat. Neuchâtel,* Vol. II.
Neuchâtel, 1839.

1839. — **56.** *Pterygotus Problematicus, onchus Murchisoni,* in
"The Silurian System," by Roderick Impey Murchison, p. 606.
4to.   London, 1839.

1839. — **57.** Mémoire sur les moules de Mollusques vivans et fossiles. Première partie. Moules d'Acéphales vivans. *Mém. Soc. sc. nat. Neuchâtel,* Vol. II., 50 pages, 10 plates. Neuchâtel, 1839. *Bull. Soc. imp. natur. Moscou,* pp. 415–430. Moscou, 1839.

1839. — **58.** Geologie und Mineralogie in Beziehung zur natür-lichen Theologie von W. Buckland, aus dem Englischen übersetzt und mit Anmerkungen und Zuzätzen versehen, von L. Agassiz. 2 vols., 80 plates. Neuchâtel, 1839.

1839. — **59.** Lettre écrite par M. Ls. Agassiz à M. Ed. Charles-worth en réponse à un article inséré dans le N° 29 du. *Magazine of Natural History,* 3 pages, 4to, autographiée. Neuchâtel, 15 mai, 1839. Reprinted under the title, "Letters from Professor Agassiz, on the subject of the French edition of the 'Mineral Conchology of Great Britain,'" Neuchâtel, 15 mai, 1839, in French first, and followed by an English translation, in *Mag. Nat. Hist.,* new series, Vol. III., N° 31, pp. 356–359. London, July, 1839.

1839. — **60.** Anhang, added at the end of a memoir by J. J. Tschudi, entitled, "Classification der Batrachier, mit berücksichti-gung der fossilen Thiere dieser Abtheilung der Reptilien," *Mém. Soc. sc. nat. Neuchâtel,* Vol. II. 4to. An appendix of two pages, dated September, 1838, and signed, Dr. Agassiz. Neuchâtel, 1839.

1839. — **61.** Observations sur les échinodermes fossiles des ter-rains de la Suisse. *Verhandl. Schweiz. naturforsch. Gesellschaft; Bern,* 5, 6, 7 *Aug.,* 1839, pp. 43–44. Bern, 1839.

1839. — **62.** Catalogus Echinodermatum fossilium musei neoco-mensis. *Bull. Soc. imp. natur. Moscou,* pp. 422–430. Moscou, 1839.

1839–1840. — **63.** Description des Échinodermes fossiles de la Suisse. Première partie, Spatangoides et Clypeastroides, 101 pages, 14 plates, 1839. Seconde partie, Cidarides, 107 pages, 11 plates, 1840. *Nouv. Mém. Soc. Helvétique sc. nat.,* Vols. III. and IV., Neuchâtel, 1839 and 1840. *Leonhard und Bronn, Neues Jahrb.,* 1840, p. 502, and 1842, p. 393.

1840. — **64.** Gletscher-Studien mit Studer. *Leonhard und Bronn, Neues Jahrb.,* pp. 92–93. 1840.

1840. — **65.** Études sur les glaciers. 1 vol., 8vo, and an atlas of 32 plates, folio. Neuchâtel, 1840.

1840. — **66.** Untersuchungen über die Gletscher; a German edition of the "Études sur les glaciers." 1 vol., 8vo, and atlas of 32 plates, folio. Neuchâtel, 1840.

1840. — **67.** Énumération des Poissons fossiles d'Italie. *Nuovi Annali Sc. Nat.*, Vol. IV., pp. 244-245 and 325-332. Bologna, 1840.

1840. — **68.** On glaciers and boulders in Switzerland. *Report British Assoc. Adv. Sc. Glasgow*, Part II., pp. 113-114. Glasgow, 1840.

1840. — **69.** On animals found in red snow. *Report British Assoc. Adv. Sc. Glasgow*, Part II., p. 143. Glasgow, 1840. *Leonhard und Bronn, Neues Jahrb.*, p. 93, 1840, under the title, "Färbende Infusorien in rothem Schnee." *L'Institut*, Vol. IX., p. 94, Paris, 1841, under the title, "Sur les animaux de la neige rouge."

1840. — **70.** On the polished and striated surfaces of the rocks which form the beds of Glaciers in the Alps. *Proc. Geol. Soc. London*, Vol. III., No. 71, June 10, 1840, pp. 321-322. *Ann. Mag. Nat. Hist.*, Vol. VI., pp. 392-393. *Edinb. New Philos. Mag.* Vol. XVIII., pp. 565-569. 1842.

1840. — **71.** On Glaciers, and the evidence of their having once existed in Scotland, Ireland, and England. *Proc. Geol. Soc. London*, Vol. III., No. 72, Nov. 4, 1840, pp. 327-332. *Ann. Mag. Nat. Hist.*, Vol. VI., pp. 396-397. 1841. *Edinb. New Phil. Mag.*, Vol. XVIII., pp. 569-570. 1842.

1840. — **72.** Observations sur la structure des écailles de poissons. *Ann. Sc. Nat. Zoologie*, 2d Série, Vol. XIII., pp. 58-61. Paris, 1840.

1840. — **73.** Remarques à l'occasion d'une note de M. Mandt sur la structure des écailles de poissons. *Comptes Rendus Acad. sc. France*, Vol. X., pp. 191-194. Paris, 1840. Reprinted under the title, "Observations sur la structure et le mode d'accroissement des écailles des poissons et réfutation des objections de M. Mandt. *Ann. Sc. Nat. Zoologie*, 2d Série, Vol. XIV., pp. 97-110. Paris, 1840. Translated into English: *Edinb. New Phil. Journ.*, Vol. XXVIII., pp. 287-291. 1840.

1840. — 74. Catalogus systematicus ectyporum echinodermatum fossilium musei neocomensis, secundum ordinem zoologicum dispositus; adjectis synonimis recentioribus, nec non stratis et locis in quibus reperiuntur. Sequuntur characteres diagnostici generum novorum vel minus cognitorum. 4to. Neocomi, Helvetorum, 1840.

1840. — 75. Études critiques sur les Mollusques fossiles. Mémoire sur les Trigonies. 58 pages. - 11 plates. 4to. Neuchâtel, 1840. *Leonhard und Bronn, Neues Jahrb.*, 1841, p. 848.

1840. — 76. Gegen Wismann's Ansricht vom Ursprung erratischer Blöcke. *Leonhard und Bronn, Neues Jahrb.*, 1840, pp. 575–576.

1841. — 77. On the fossil fishes found by Mr. Gardner in the province of Ceará, in the north of Brazil. *Edinb. New Phil. Journ.*, Vol. XXX., pp. 82–84. ˙1841.

1841. — 78. Genus *Trigonia.* Character von Artüberhaupt Gletscher. *Leonhard und Bronn, Neues Jahrb.*, 1841, pp. 356–357.

1841. — 79. Alter Moränen bei Baden-Baden. *Leonhard und Bronn, Neues Jahrb.*, 1841, pp. 566–567.

1841. — 80. Une série de coquilles vivantes et fossiles des bords de la Clyde en Ecosse. *Verhandl. Schweiz. naturforsch. Gesellschaft*; *Zürich*, 2, 3, 4 *Aug.*, 1841, pp. 63–64. Zürich, 1841.

1841. — 81. Routes parallèles de Glen-Roy en Ecosse. *Verhandl. Schweiz. naturforsch. Gesellschaft*; *Zürich*, 2, 3, 4 *Aug.*, 1841, pp. 68, 69. Zürich, 1841.

1841. — 82. Terrain cyliolitique. *Verhandl. Schweiz. naturforsch. Gesellschaft*; *Zürich*, 2, 3, 4 *Aug.*, 1841, p. 72. Zürich, 1841.

1841. — 83. Monographies d'Échinodermes, vivans et fossiles. 2ième livraison; "les Scutelles." En tête de cette livraison, se trouve: "Observations sur les progrès récents de l'histoire naturelle des Échinodermes." 4to. Neuchâtel, 1841.

Nota bene. — Les "Observations sur les progrès, etc.," ont été tirés à part et distribués en juillet, 1841. 20 pages, 4to, avec une page de title: "Monograph d'Échinodermes (Extrait de la seconde livraison de cet ouvrage, etc.)." English translation: "Observations on the progress recently made in the natural history of the echinodermata."

Ann. Mag. Nat. Hist., Vol. IX., pp. 180-190 and 569-570. London, 1842.

1841.—**84.** Additions to Mr. Wood's catalogue of Crag radiaria. *Ann. Mag. Nat. Hist.*, Vol. VI., p. 343. London, 1841.

1841.—**85.** De la succession et du développement des êtres organisés à la surface du globe terrestre dans les différents âges de la nature. Discours prononcé à l'inauguration de l'Académie de Neuchâtel, le 18 novembre, 1841. 47 pages. 12mo. Neuchâtel, 1841. The edition was limited to four hundred copies, and two hundred separata for the use of Agassiz. Translated into English: *Edinb. New Phil. Journ.*, Vol. XXXIII., pp. 388-399. Edinburgh, October, 1842. A German translation by Dr. N. Gräger was published at Halle, 1843. Part of it is reprinted in *La première Académie de Neuchâtel*, par A. Petitpierre, pp. 85-93. Neuchâtel, 1889.

1842.—**86.** The Prospectus of the Nomenclator Zoologicus; in French and in German. 3 pages. 4to. Neuchâtel and Soleure, février, 1842.

1842.—**87.** Nomenclator Zoologicus, continens nomina systematica generum animalium tam viventium quam fossilium, etc. 4to. Soloduri, 1842. *Leonhard und Bronn, Neues Jahrb.*, 1842, p. 496.

1842.—**88.** Deux Lettres à M. Arago. Observations sur le glacier de l'Aar et sur les glaciers. *Comptes Rendus Acad. sc. France*, Vol. XV., pp. 284-288 and 435-466. Paris, 1842. *L'Institut*, Vol. X., pp. 278, 305, and 359. Paris, 1842. *Leonhard und Bronn, Neues Jahrb.*, 1843, p. 364. *Edinb. New Phil. Journ.*, Vol. XXXIII., p. 339. Edinburgh, 1842.

1842.—**89.** La théorie des glaces et ses progrès les plus récents. *Bibl. Univ. Genève*, Vol. XLI., pp. 118-139. Genève, 1842. *Leonhard und Bronn, Neues Jahrb.*, 1842, pp. 56-58. *Edinb. New Phil. Journ.*, Vol. XXXIII., p. 217, and Vol. XXXIV., p. 364, under the title: "The glacial theory and its recent progress."

1842.—**90.** Histoire naturelle des poissons d'eau douce de l'Europe centrale. 1 vol., 8vo, and folio atlas of 41 coloured plates. Neuchâtel, 1842.

1842. — **91.** Monographies d'Échinodermes, vivans et fossiles. 4ième livraison, "L'anatomie du genre echinus," par G. Valentin; avec une "Preface par Agassiz." 4to. Neuchâtel, 1842.

The "preface" of ten pages by Agassiz is important because in it he says that Valentin delivered his manuscript into his hands as far back as 1840, and the "Preface" is dated December, 1841. The exact date of the issue of the fourth part or "livraison" is marked on the cover, — 1842. It is the last part of those monographs of the Echinodermata.

The third monograph, "Des Galérites," and the fourth monograph, "Des Dysaster," forming together the third part, or "3ième livraison," are by E. Desor; and the part was issued in 1842. Like the other parts, it was published "aux frais de L. Agassiz."

1842. — **92.** No title. The paper has been quoted under the following designations: 1st, "Lettre sur la structure lamellaire des glaciers que s'attribue Mr. James D. Forbes." 2d, "A reply to Mr. James D. Forbes on the laminated structure of glaciers." 3d, "Agassiz's controversy with James D. Forbes, of Edinburgh." Neuchâtel, 10 pages. 4to. 11 avril, 1842. Expédié de Neuchâtel, le 21 avril, 1842. In it is reprinted a letter of Professor James D. Forbes to M. Desor, which was largely distributed among savants by Professor Forbes.

1842. — **93.** Extrait d'une lettre à M. Alcide d'Orbigny sur la répartition des fossiles dans les divers terrains. *Bull. Soc. Géol. France*, Vol. XIII., pp. 355–356. Paris, mai, 1842.

1842. — **94.** New Views regarding the distribution of fossils in formations. *Edinb. New Phil. Journ.*, Vol. XXXII., pp. 97–98. Edinburgh, January, 1842.

1842. — **95.** Matériaux pour une bibliothèque zoologique et paléontologique. 562 pages, folio. Neuchâtel, 1842.

1842. — **96.** Reiseproject nach dem Aargletscher, Hügi über Gletscher; Myaceen. *Leonhard und Bronn, Neues Jahrb.*, pp. 313–317. 1842.

1842. — **97.** Vortrag über seine Gletscheruntersuchungen auf dem Aargletscher. *Verhandl. Schweiz. Naturforsch. Gesellschaft ; Altdorf,* 25, 26, *und* 27 *Juli,* 1842, pp. 81–91. Altdorf, 1842.

1842. — 98. Betreff der Gletschereinwirkung im Jura. *Verhandl. Schweiz. Naturforsch. Gesellschaft ; Altdorf,* 25, 26, *und* 27 *Juli,* 1842, pp. 46–48. Altdorf, 1842.

1842. — 99. Erwiederung auf Dr. Carl Schimper's Angriffe. 4 pages, 4to. (Im November, 1842, privatim vertheilt.) No place of publication, although there is no doubt that it was Neuchâtel.

1842–1845. — 100. Études critiques sur les mollusques fossiles. Monographie des Myes. 287 pages. 94 plates. 4to. Neuchâtel, 1842–1845. *Leonhardt und Bronn, Neues Jahrb.,* 1842, p. 862, and 1843, p. 747.

1843. — 101. Neue Beobachtungen auf den Gletschern; Myaceen; Struktur der Gletscher; Desor über fossile Nucleoliten; Fossilarten der Molasse. *Leonhard und Bronn, Neues Jahrb.,* 1843, pp. 84–89.

1843. — 102. Synoptical table of British fossil fishes, arranged in the order of geological formation. *Report British Assoc. Adv. Sc.* 1843, pp. 194–207. *Edinb. New Phil. Journ.,* Vol. XXXVII., pp. 331–347. *Ann. Sc. Nat. Zoologie,* 3ᵉ Série, pp. 251–271. Paris, 1844.

1843. — 103. A period in the history of our planet. *Edinb. New Phil. Journ.,* Vol. XXXV., pp. 1–291. Edinburgh, 1843.

1843. — 104. Notice sur la succession des Poissons fossiles dans la série des formations géologiques. (Extrait de la 18ᵉ et dernière livraison des *Recherches sur les Poissons fossiles.*) 16 pages, 4to. Neuchâtel, 1843. *Ann. Sc. Nat. Zoologie,* 3ᵉ Série, Vol. II., pp. 251–271. Paris, 1844.

1843. — 105. Sur les glaciers. *Actes Soc. Helvétique sc. nat. Lausanne,* 24, 25, *et* 26 *juillet,* 1843, pp. 72–74. Lausanne, 1843.

1843. — 106. Valeur géologique des dents de Squales pour la détermination des terrains. *Actes Soc. Helvétique sc. nat. Lausanne,* 24, 25, *et* 26 *juillet,* 1843, pp. 83–84. Lausanne, 1843.

1843. — 107. Quel est l'âge des plus grands glaciers des Alpes suisses? Lettre à M. Arago. *Comptes Rendus Acad. sc. France,* Vol. XVI., p. 678. Paris, 1843.

1843. — 108. Sur le Mouvement du glacier de l'Aar. Influence de l'inclinaison du sol sur le mouvement de la glace. *Bull. Soc. sc. nat. Neuchâtel,* Vol. I., pp. 1–5. Neuchâtel, novembre, 1843. *Leonhard und Bronn, Neues Jahrb.,* p. 620. 1844.

1843. — 109. Sur les fossiles rapportés du Pérou par M. Tschudi. *Bull. Soc. sc. nat. Neuchâtel,* Vol. I., pp. 29–30. Neuchâtel, decembre, 1843.

1843. — 110. Sur la détermination exacte de la limite des neiges éternelles en un point donné. *Comptes Rendus Acad. sc. France,* Vol. XVI., p. 752. *Poggend. Ann. der Phys. und Chem.,* Vol. LIX., p. 342.

1843. — 111. Exposé abrégé de la stratification des glaciers avec des coupes. *Leonhard und Bronn, Neues Jahrb.,* 1843, pp. 84 and 86.

1843. — 112. Report on the fossil fishes of the Devonian system or old red sandstone. *Report British Assc. Adv. Sc.* 1842, pp. 80–88. *Bibl. univ. Genève,* Vol. XLIII., pp. 353–369. 1843.

1844. — 113. Tableau général des Poissons fossiles rangés par terrains. 17 pages. 4to. Neuchâtel, 1844. (Extrait des *Recherches sur les Poissons fossiles.*) Introduction; Neuchâtel, 1844. *Ann. sc. nat. Zoologie,* 3ᵉ série, Vol. II., pp. 251–271. Paris, 1844.

1844. — 114. Essai sur la classification des Poissons. (Extrait de la 18ᵉ et dernière livraison des *Recherches sur les Poissons fossiles.*) 8 pages and a table. 4to. Neuchâtel, 1844. *Edinb. New Phil. Journ.,* Vol. XXXVII., pp. 132–143. 1844.

1844. — 115. De la forme des Placoïdes et Tableau général des espèces de placoïdes rangés par terrains. (Extrait de la dernière livraison des *Recherches sur les Poissons fossiles.*) 18 pages. 4to. Neuchâtel, 1844.

1844. — 116. Sur les progrès de l'étude de l'ichthyologie. *Bull. Soc. sc. nat. Neuchâtel,* Vol. I., pp. 49, 50. Neuchâtel, janvier, 1844.

1844. — **117.** Sur l'importance des divers embranchemens du règne animal, sous le point de vue biologique, ou revue des différentes époques géologiques. *Bull. Soc. sc. nat. Neuchâtel*, Vol. I., pp. 50–52. Neuchâtel, janvier, 1844.

1844. — **118.** Sur l'Isar des Pyrénées, comparé au chamois des Alpes. *Bull. Soc. sc. nat. Neuchâtel*, Vol. I., pp. 57–58. Neuchâtel, février, 1844.

1844. — **119.** Sur la distribution géographique des quadrumanes. *Bull. Soc. sc. nat. Neuchâtel*, Vol. I., pp. 59–62. Neuchâtel, mars, 1844.

1844. — **120.** Sur la distribution géographique des Chiroptères. *Bull. Soc. sc. nat. Neuchâtel*, Vol. I., pp. 63–65. Neuchâtel, mars, 1844.

1844. — **121.** Sur le genre *Pyrula* de Lamarck. *Bull. Soc. sc. nat. Neuchâtel*, Vol. I., pp. 69–70. Neuchâtel, avril, 1844.

1844. — **122.** Sur les prétendues identités que l'on admet généralement entre les espèces vivantes et les fossiles de certaines terrains. *Bull. Soc. sc. nat. Neuchâtel*, Vol. I., pp. 107, 108. Neuchâtel, mai, 1844.

1844. — **123.** On fossil fishes. *Edinb. New Phil. Journ.*, Vol. XXXVII., pp. 331–334. 1844.

1844. — **124.** Sur quelques Poissons fossiles du Brésil. *Comptes Rendus Acad. sc. France*, Vol. XVIII., pp. 1007–1015. Paris, 1844. *L'Institut*, Vol. XII., pp. 187–188. Paris, 1844.

1844. — **125.** Réponse à la société académique de Savoie. *Bull. Soc. géol. France*, 2ᵉ série, Vol. I., pp. 605–606. Chambéry, août, 1844.

1844. — **126.** Sur un nouvel oursin, le *Metaporinus Michelini*. *Bull. Soc. géol. France*, 2ᵉ série, Vol. I., p. 730. Chambéry, août, 1844.

1844. — **127.** Sur les prétendues identités des coquilles tertiaires et vivantes. *Bull. Soc. géol. France*, 2ᵉ série, Vol. I., pp. 744–745. Chambéry, août, 1844.

1844.— **128.** Observations sur le poli et les stries des roches. *Bull. Soc. géol. France*, 2ᵉ série, Vol. I., pp. 617, 630, 639, 640. Chambéry, août, 1844.

1844.— **129.** Sur les glaciers et les dépôts erratiques. *Bull. Soc. géol. France*, 2ᵉ série, Vol. I., pp. 636, 637, 643, 646, 648, 650. Chambéry, août, 1844.

1844.— **130.** Nummulitique dans la craie supérieur. *Bull. Soc. géol. France*, 2ᵉ série, Vol. I., pp. 826, 827, 629, 630. Chambéry, août, 1844.

1844.— **131.** La question paléontologique de Petit-Cœur. *Bull. Soc. géol. France*, 2ᵉ série, Vol. I., p. 672. Chambéry, août, 1844.

1844.— **132.** Sur une collection de coquilles d'Orient de M. Albert de Pourtalès. *Bull. Soc. sc. nat. Neuchâtel*, Vol. I., pp. 141, 142. Neuchâtel, novembre, 1844.

1844.— **133.** Sur le cerveau des Poissons. *Bull. Soc. sc. nat. Neuchâtel*, Vol. I., pp. 147-148. Neuchâtel, décembre, 1844.

1844.— **134.** Introduction à une monographie des poissons fossiles du vieux grès rouge; suivi d'un tableau synoptique des poissons fossiles du système Dévonien. (Extrait de la *Monographie des Poissons fossiles de l'Old Red Sandstone ou système Dévonien*.) 36 pages. 4to. Soleure, 1844.

1844-1845.— **135.** Monographies des Poissons fossiles, supplément aux Recherches sur les Poissons fossiles. Première monographie, comprenant l'histoire des Poissons du vieux grès Rouge (Old Red Sandstone ou système Dévonien). Trois livraisons ou parts, in 4to, avec atlas in folio de 43 planches. Neuchâtel, 1844-45.

1845.— **136.** Lettres sur les poissons fossiles du système Dévonien de la Russie, adressées à MM. Murchison et de Verneuil par M. le professeur Agassiz. *Russia and the Ural Mountains* or *Géologie de la Russie d'Europe et les Montagnes de l'Oural*, par Roderick Impey Murchison, Edouard de Verneuil, Alexandre de Keyserling, Vol. II., *Paléontologie* ; Appendice, pp. 397-418. Five letters written from 1842 to 1845. 4to. London and Paris.

1845.— **137.** Iconographie des coquilles tertiaires réputées identiques avec les espèces vivantes. *Nouv. Mém. Soc. Helvétique Sc. Nat.*, Vol. VII. Neuchâtel, 1845.

1845. — **138.** Sur les métamorphoses des animaux des classes inférieures. *Bull. Soc. sc. nat. Neuchâtel*, Vol. I., pp. 156–159. Neuchâtel, janvier, 1845.

1845. — **139.** Sur la distribution géographique des animaux et de l'homme. *Bull. Soc. sc. nat. Neuchâtel*, Vol. I., pp. 162–166. Neuchâtel, janvier, 1845.

1845. — **140.** Cirques des glaciers dans les Alpes et le Jura. *Bull. Soc. sc. nat. Neuchâtel*, Vol. I., p. 172. Neuchâtel, février, 1845.

1845. — **141.** Anciennes moraines de l'Allée-Blanche et du val Ferret. Glacier d'Ornex. *Bull. Soc. sc. nat. Neuchâtel*, Vol. I., p. 171. Neuchâtel, février, 1845.

1845. — **142.** Lettre à Élie de Beaumont sur les roches striées de la Suisse. *Bull. Soc. géol. France*, 2ᵉ série, Vol. II., pp. 273–277. Paris, février. 1845.

1845. — **143.** Sur un fait de superposition de roches observées en Ecosse par M. Robertson. *Bull. Soc. sc. nat. Neuchâtel*, Vol. I., pp. 183–184. Neuchâtel, mars, 1845.

1845. — **144.** Sur l'importance de l'étude des animaux fossiles. *Bull. Soc. sc. nat. Neuchâtel*, Vol. I., pp. 189–190. Neuchâtel, avril, 1845.

1845. — **145.** Extrait de deux lettres de L. Agassiz à Edouard Collomb sur des galets striés glaciaires des Vosges. *Bull. Soc. géol. France*, 2ᵉ série, Vol. II., pp. 394–395. Paris, mai, 1845.

1845. — **146.** Anatomie des Salmones par L. Agassiz et C. Vogt. 14 planches. 4to. *Mém. Soc. sc. nat. Neuchâtel*, Vol. III. Neuchâtel. 1845.

1845. — **147.** Remarques sur le " Traité de Paléontologie " de F. Jules Pictet. *Bibl. univ. Genève*, N° 112, juin, 1845. *Edinb. New Phil. Journ.*, Vol. XXXIX., pp. 295–302. 1845.

1845. — **148.** Notice sur la géographie des animaux. *Revue Suisse*, p. 31. Neuchâtel, août, 1845.

1845. — **149.** Nouvelles observations sur les nageoires des.poissons. *Actes Soc. Helvétique sc. nat. Genève*, 11, 12 *et* 13 *août*, 1845, p. 49. Genève, 1845.

1845. — **150.** Observations sur le glacier de l'Aar. *Actes Soc. Helvétique sc. nat. Genève*, 11, 12 *et* 13 *août*, 1845, p. 66. Genève, 1845.

1845. — **151.** Observations du Dr. Basswitz sur la neige rouge. *Actes Soc. Helvétique sc. nat. Genève*, 11, 12 *et* 13 *août*, 1845, p. 68. Genève, 1845.

1845. — **152.** Résumé de ses travaux sur l'encéphale des poissons. *Actes Soc. Helvétique sc. nat. Genève*, 11, 12 *et* 13 *août*, 1845, p. 70. Genève, 1845.

1845. — **153.** Sur diverses familles de l'ordre des crinoides. *Actes Soc. Helvétique sc. nat. Genève*, 11, 12 *et* 13 *août*, 1845, pp. 91–92. Genève, 1845.

1845. — **154.** Notice sur les glaciers de l'allée Blanche et du Val-Ferret. In *Nouvelles Excursions et séjours dans les glaciers et les hautes régions des Alpes, de M. Agassiz et ses compagnons de voyage,* par E. Desor, pp. 212–219. Neuchâtel, 1845. In the table of contents, and on page 212 of the book, the title is different from the title on the cover and the title-page; it there reads : "Les glaciers et le terrain erratique du revers méridional du Mont Blanc, par M. Agassiz."

1845. — **155.** Remarques sur les observations de M. Durocher relatives aux phénomènes erratiques de la Scandinavie. *Comptes Rendus Acad. sc. France,* Vol. XXII., pp. 1331–1333. Paris, 1845. *Edinb. New Phil. Journ.*, Vol. XL., p. 237. 1845.

1845. — **156.** Rapport sur les poissons fossiles de Pargile de Londres. — Report on the fossil fishes of the London clay (French and English). *Report British Assoc. Adv. Sc.*, 1844, pp. 279–310, one plate. *Ann. sc. nat. Zoologie*, pp. 21–48. Paris, 1845. *Edinb. New Phil. Journ.*, Vol. XXXIX., pp. 321–327.

1845–1846. — **157.** On fossil fishes, particularly those of the London clay. *Edinb. New Phil. Journ.*, Vol. XXXIX., pp. 321–327, 1845 ; and Vol. XL., pp. 121–125, 1846.

1846. — **158.** Nomenclatoris Zoologici Index universalis, etc. 4to. Soloduri, 1846. A 12mo edition of this work was issued also at Soloduri, in 1848, by the same publishers, Jent et Gasmann.

1846. — **159.** Observations sur la distribution géographique des êtres organisés. *Bull. Soc. sc. nat. Neuchâtel*, Vol. I., pp. 357–362. Neuchâtel, 1846.

1846. — **160.** Observations sur les rapports qui existent entre les faits relatifs à l'apparition successive des êtres organisés à la surface du globe et la distribution géographique des différents types actuels d'animaux. *Bull. Soc. sc. nat. Neuchâtel*, Vol. I., pp. 366–369. Neuchâtel, 1846.

1846. — **161.** Observations sur les glaciers de la Suisse. *Bull. Soc. géol. France*, 2ᵉ série, Vol. III., pp. 415–418. Paris, avril, 1846.

1846. — **162.** Discussions sur l'Oscillation des glaciers et sur les roches et galets striés. *Bull. Soc. géol. France*, 2ᵉ série, Vol. III., pp. 419–422. Paris, avril, 1846.

1846. — **163.** Sur un nouveau genre de poissons fossiles (*Emidichthys*) du terrain dévonien de l'Eifel. *Bull. Soc. géol. France*, 2ᵉ série, Vol. III., pp. 488–489. Paris, mai, 1846.

1846. — **164.** Sur les Poissons des terrains paléozoiques. *Ext. Proc. Verb. Soc. Philomathique*, pp. 61–62. Paris, 1846. *L'Institut*, Vol. XIV., p. 163. Paris, 1846.

1846. — **165.** On the ichthyological fossil fauna of the Old Red Sandstone. *Edinb. New Phil. Journ.*, Vol. XLI., pp. 17–49. 1846.

1846. — **166.** Résumé d'un travail d'ensemble sur l'organisation, la classification et le développement progressif des Échinodermes dans la série des terrains. *Comptes Rendus Acad. sc. France*, Vol. XXXIII., pp. 276–279. Paris, 1846.

1846-47-48. — **167.** Catalogue raisonné des familles, des genres et des espèces de la classe des Échinodermes, par L. Agassiz et E. Desor; précédé d'une introduction sur l'organisation, la classification et le développement progressif des types dans la série des terrains, par. L. Agassiz. *Ann. sc. nat. Zoologie*, 3ᵉ série. Vol. VI., pp. 350-374, 1846; Vol. VII., pp. 129-168, 1847; Vol. VIII., pp. 5–35, et pp. 355–381, 1848, Paris. Sixty separates printed with repaging; 167 pages, 2 plates. Paris, 1848.

1846. — **168.** On the fish *Huronigricans* of Cuvier; and the shovel fish from the Ohio River. *Proc. Boston Soc. Nat. Hist.,* Vol. II., p. 184. Boston, November, 1846.

1846. — **169.** On a new *Pygorhynchus* of Georgia; and the crawfish from the Mammoth Cave, Kentucky. *Proc. Boston Soc. Nat. Hist.,* Vol. II., pp. 193–194. Boston, December, 1846.

1846. — **170.** Climate of Europe during the later Miocene period. *Proc. Amer. Acad. Arts and Sc.,* Vol. I., p. 43. Boston, December, 1846.

1847. — **171.** Analogy between the fossil flora of the European Miocene and the living flora of America, in a letter to Roderick I. Murchison. *Athenæum,* No. 1023. London, 1847. Reproduced in *Amer. Journ. Sc.,* 2d series, Vol. IV., pp. 424–425. 1847.

1847. — **172.** On the Coal-field of Eastern Virginia — Fossil Fishes, by Charles Lyell, with notes from L. Agassiz and Sir Philip Egerton. *Quart. Journ. Geol. Soc. London,* Vol. III., pp. 275–278. London.

1847. — **173.** Geologische Alpenreisen. L. Agassiz. Deutsche von Desor und Vogt. Zweite Aufl. mit. 4 Karten. Frankfurt, 1847.

1847. — **174.** Système glaciaire ou Recherches sur les glaciers, leur mécanisme, leur ancienne extension et le rôle qu'ils ont joué dans l'histoire de la terre, par MM. L. Agassiz, A. Guyot et E. Desor. Première partie : Nouvelles études et expériences sur les glaciers actuels, leur structure, leur progression et leur action physique sur le sol ; par L. Agassiz. 1 vol., 8vo, et un atlas in folio, avec 3 cartes et 9 planches. Paris et Neuchâtel, mai, 1847. The other two parts, by Guyot and Desor, were never published, or even written.

1847. — **175.** Lettre à M. de Humboldt sur quelques points de l'organisation des animaux Rayonnés et sur la parité bilatérale dans les Actinies. *Comptes Rendus Acad. sc. France,* Vol. XXV., pp. 677–682.　Paris, 1847.　*Proc. Verb. Soc. Philomathique,* pp. 95–98. Paris, 1847.　*L'Institut,* Vol. XV., pp. 388–389.　Paris, 1847.　*Edinb. New Phil. Journ.,* Vol. XLIV., pp. 316–319.　Edinburgh, April, 1848.

1847. — **176.** An introduction to the study of natural history, in a series of lectures delivered in the hall of the College of Physicians and Surgeons, New York; also a biographical notice of the author. 58 pages, 8vo.　New York, 1847.

1847. — **177.** Remarks on the position of Boston naturalists in their location at a seaport.　*Proc. Boston Soc. Nat. Hist.,* Vol. II., p. 243.　Boston, July, 1847.

1847. — **178.** On the blind-fish of the Mammoth Cave.　*Proc. Amer. Acad. Arts and Sc.,* Vol. I., p. 180.　Boston, October, 1847.

1847. — **179.** Remarks upon the Moose and Caribou (*Cervus alces et Tarandus*) and the American Raven.　*Proc. Boston Soc. Nat. Hist.,* Vol. II., pp. 187, 188.　Boston, November, 1847.

1847. — **180.** De l'étude comparative des animaux inférieurs et des plantes qui accompagnent l'homme en Europe et dans l'Amérique.　*Bull. Soc. sc. nat. Neuchâtel,* Vol. II., pp. 187–189.　Neuchâtel, December, 1847.

1848. — **181.** Letter to Dr. Gibbes in relation to *Dorudon Serratus.　Proc. Acad. Nat. Sc. Philadelphia,* Vol. IV., pp. 4–5. Philadelphia, February, 1848.

1848. — **182.** Tubulibranchiate Annelids of Boston harbor.　*Proc. Boston Soc. Nat. Hist.,* Vol. III., pp. 26–27.　Boston, April, 1848.

1848. — **183.** On the existence of numerous minute tubes in Fishes.　*Proc. Boston Soc. Nat. Hist.,* Vol. III., pp. 27–28.　Boston, April, 1848.

1848. — **184.** An appeal to the students of science in America. *Proc. Boston Soc. Nat. Hist.,* Vol. III., pp. 36–37.　Boston, May, 1848.

1848. — **185.** Observations on the structure of the foot in the embryo of birds.　*Proc. Boston Soc. Nat. Hist.,* Vol. III., pp. 42–43.　Boston, June, 1848.

1848. — **186.** The terraces and ancient river bars, drifts, boulders, and polished surfaces of Lake Superior. *Proc. Amer. Assoc. Adv. Sc.*, First Meeting, Philadelphia, pp. 68–70. Philadelphia, 1849.

1848. — **187.** Fishes of Lake Superior. *Proc. Amer. Assoc. Adv. Sc.*, First Meeting, Philadelphia, pp. 30–32. Philadelphia, 1849.

1848. — **188.** On the comparison of Alpine and northern vegetation. *Proc. Amer. Assoc. Adv. Sc.*, First Meeting, Philadelphia, pp. 41–42. Philadelphia, 1849.

1848. — **189.** Phonetic apparatus of the Cricket. *Proc. Amer. Assoc. Adv. Sc.*, First Meeting, Philadelphia, p. 41. Philadelphia, 1849.

1848. — **190.** Black Banded Cyprinidæ. *Proc. Amer. Assoc. Adv. Sc.*, First Meeting, Philadelphia, p. 70. Philadelphia, 1849.

1848. — **191.** Monograph of Garpikes. *Proc. Amer. Assoc. Adv. Sc.*, First Meeting, Philadelphia, pp. 70–71. Philadelphia, 1849.

1848. — **192.** On the origin of the actual outlines of Lake Superior. *Proc. Amer. Assoc. Adv. Sc.*, First Meeting, Philadelphia, p. 79. Philadelphia, 1849.

1848. — **193.** On the fossil Cetacea of the United States. *Proc. Amer. Acad. Arts and Sc.*, Vol. II., pp. 4–5. Boston, October, 1848.

1848. — **194.** The *Salmonidæ* of Lake Superior. *Proc. Boston Soc. Nat. Hist.*, Vol. III., pp. 61–62. Boston, October, 1848.

1848. — **195.** Revision of the system of classification in Zoölogy. *Proc. Boston Soc. Nat. Hist.*, Vol. III., p. 65. Boston, October, 1848.

1848. — **196.** Two new Fishes from Lake Superior (*Percopsis Rhinichthys*). *Proc. Boston Soc. Nat. Hist.*, Vol. III., pp. 80–81. Boston, November, 1848.

1848. — **197.** Principles of Zoölogy: touching the structure, development, distribution, and natural arrangement of the races of animals, living and extinct. Part I., Comparative Physiology, by Louis Agassiz and Augustus A. Gould. 12mo. Boston, 1884. Second edition, 1851. Third edition, 1861. Translated into German by H. G. Bronn, with an Introduction by Bronn. 8vo. Stuttgart, 1851. Translated into French by Elisée Reclus, in *Magasin d'Éducation et de Récréation*, 1891. Published by Hetzel & Co., Paris. Several unauthorized editions were published in England and in Germany between the years 1849 and 1854. Part II., Systematic Zoölogy, although advertised, was never published.

1848–1854. — **198.** Bibliographia Zoologiæ et Geologiæ. A general catalogue of all books, tracts, and memoirs of Zoölogy and Geology. By Professor Louis Agassiz. Corrected, enlarged, and edited by H. E. Strickland and Sir William Jardine (*Ray Society*). 4 vols. London, 1848–1854.

1849. — **199.** Twelve lectures on Comparative Embryology delivered before the Lowell Institute in Boston, December and January, 1848–1849. 104 pages, 8vo. Reprinted from the *Daily Evening Traveller.* Boston, 1849.

1849. — **200.** Communication from Agassiz relative to the formation of a Museum at the Smithsonian Institution. *U. S. Public Documents*, 31st Congress, 1st Session. House and Senate Misc. Doc., pp. 24–26. Washington, 1849.

1849. — **201.** On the distinction between the fossil crocodiles of the Green Sand of New Jersey. *Proc. Acad. Nat. Sc. Philadelphia*, Vol. IV., p. 169. Philadelphia, March, 1849.

1849. — **202.** Investigations upon Medusæ. *Proc. Amer. Acad. Arts and Sc.*, Vol. II., pp. 148–149. Boston, May, 1849.

1849. — **203.** On the structure of coral animals. *Proc. Amer. Assoc. Adv. Sc.*, Second Meeting, Cambridge, August, 1849, pp. 68–77. Boston, 1850.

1849. — **204.** The zoölogical character of young mammalia. *Proc. Amer. Assoc. Adv. Sc.*, Second Meeting, Cambridge, August, 1849, pp. 85–89. Boston, 1850.

1849. — **205.** The vegetable character of *Xanthidium. Proc.*

Amer. Assoc. Adv. Sc., Second Meeting, Cambridge, August, 1849, pp. 89–91. Boston, 1850.

1849. — **206.** On the fossil remains of an elephant found in Vermont. *Proc. Amer. Assoc. Adv. Sc.*, Second Meeting, Cambridge, August, 1849, pp. 100–101. Boston, 1850.

1849. — **207.** On the circulation of the fluids in insects. *Proc. Amer. Assoc. Adv. Sc.*, Second Meeting, Cambridge, August, 1849, pp. 140–143. Boston, 1850. *Ann. Sc. Nat. Zoologie*, Vol. XV., pp. 358–362. Paris, 1851.

1849. — **208.** On the embryology of *Ascidia*, and the characteristics of new species from the shores of Massachusetts. *Proc. Amer. Assoc. Adv. Sc.*, Second Meeting, Cambridge, August, 1849, pp. 157–159. Boston, 1850.

1849. — **209.** On the structure and homologies of radiated animals, with reference to the systematic position of the Hydroid Polypi. *Proc. Amer. Assoc. Adv. Sc.*, Second Meeting, Cambridge, August, 1849, pp. 389–396. Boston, 1850.

1849. — **210.** On animal morphology. *Proc. Amer. Assoc. Adv. Sc.*, Second Meeting, Cambridge, August, 1849, pp. 411–423. Boston, 1850.

1849. — **211.** On the differences between progressive, embryonic, and prophetic types in the succession of organized beings through the whole range of geological times. *Proc. Amer. Assoc. Adv. Sc.*, Second Meeting, Cambridge, August, 1849, pp. 432–438. Boston, 1850. (The name of L. Agassiz was accidentally omitted, but is given in the Index, p. 451.) *Edinb. New Phil. Journ.*, Vol. XLIX., pp. 160–165. Edinburgh, 1850.

1849. — **212.** Remarks on two kinds of drift in Cambridge, on the road to Mount Auburn. *Proc. Boston Soc. Nat. Hist.*, Vol. III., p. 183. Boston, October, 1849.

1849. — **213.** Worms of the coast of Massachusetts. *Proc. Boston Soc. Nat. Hist.*, Vol. III., pp. 190–191. Boston, November, 1849.

1849. — **214.** The metamorphoses of the *Lepidoptera*. *Proc.*

Boston Soc. Nat. Hist., Vol. III., pp. 199–200.   Boston, November, 1849.

1849. — **215.** On the development of ova in insects.   *Proc. Amer. Acad. Arts and Sc.*, Vol. II., p. 181.   Boston, November, 1849.

1849. — **216.** Relation between the structure of animals and the element in which they dwell.   *Proc. Amer. Acad. Arts and Sc.*, Vol. II., p. 181.   Boston, November, 1849.

1849. — **217.** On the egg in vertebrate animals as a means of classification.   *Proc. Amer. Acad. Arts and Sc.*, Vol. II., pp. 183–184.   Boston, December, 1849.

1849. — **218.** On the circulation and digestion in the lower animals.   *Proc. Boston Soc. Nat. Hist.*, Vol. III., pp. 206–207.   Boston, December, 1849.

1849. — **219.** Resemblance of the mastodon and the manatee.   *Proc. Boston Soc. Nat. Hist.*, Vol. III., p. 209.   Boston, December, 1849.

1849. — **220.** On the respiratory system in the lower animals.   *Proc. Boston Soc. Nat. Hist.*, Vol. III., pp. 209–210.   Boston, December, 1849.

1850. — **221.** Lake Superior; its physical character, vegetation, and animals, compared with those of other and similar regions. With a narrative of the tour by J. Elliot Cabot.   Contributions by other scientific gentlemen (John L. Leconte, A. A. Gould, J. E. Cabot, T. W. Harris, A. Gray, Leo Lesquereux and Edward Tuckerman).   8vo.   Boston, 1850.

1850. — **222.** Contributions to the natural history of the Acalephæ of North America.   Part I.: On the naked-eyed Medusæ of the shores of Massachusetts in the perfect state of development. Part II.: On the Beroid Medusæ of the shores of Massachusetts in their perfect state of development.   *Mem. Amer. Acad. Arts and Sc.*, New Series, Vol. IV., pp. 221–374.   16 plates.   Boston, 1850. Communicated to the Academy May 8 and May 29, 1849.   Review of Part I. in *Amer. Journ. Sc.*, Vol. X., pp. 272–276, September, 1850.

1850. — **223.** *Phocœna Americana*, New Sp. Agas. *Proc. Boston Soc. Nat. Hist.*, Vol. III., p. 225. Boston, January, 1850.

1850. — **224.** On the gills of Crustacea. *Proc. Boston Soc. Nat. Hist.*, Vol. III., pp. 225-226. Boston, January, 1850.

1850. — **225.** Muscular structure of Medusæ. *Proc. Boston Soc. Nat. Hist.*, Vol. III., p. 232. Boston, January, 1850.

1850. — **226.** Embryonic development of insects. *Proc. Boston Soc. Nat. Hist.*, Vol. III., pp. 236-237. Boston, January, 1850.

1850. — **227.** Breathing organs of Mollusks. *Proc. Boston Soc. Nat. Hist.*, Vol. III., p. 237. Boston, January, 1850.

1850. — **228.** Remarks on the development of air-bladders. *Proc. Amer. Assoc. Adv. Sc.*, Third Meeting, Charleston, p. 72. Charleston, 1850.

1850. — **229.** Remarks on the species common to different formations. *Proc. Amer. Assoc. Adv. Sc.*, Third Meeting, Charleston, p. 73. Charleston, 1850.

1850. — **230.** On the morphology of the Medusæ. *Proc. Amer. Assoc. Adv. Sc.*, Third Meeting, Charleston, pp. 119-122. Charleston, 1850. *Edinb. New Phil. Journ.*, Vol. L., pp. 85-89. Edinburgh, 1851.

1850. — **231.** On the principles of classification (of the animal kingdom). *Proc. Amer. Assoc. Adv. Sc.*, Third Meeting, Charleston, pp. 89-96. Charleston, 1850.

1850. — **232.** On the structure of the Halcyonoid Polypi. *Proc. Amer. Assoc. Adv. Sc.*, Third Meeting, Charleston, pp. 207-213. Charleston, 1850.

1850. — **233.** Geographical distribution of animals. *Christian Examiner and Religious Miscellany.* Vol. XLVIII., pp. 181-204. Boston, March, 1850. A short résumé is published in *Bull. Soc. sc. nat. Neuchâtel*, Vol. II., février, 1852, pp. 347-350, par L. Coulon père. Neuchâtel, 1852. Reprinted in *Edinb. New Phil. Journ.*, Vol. XLIX., pp. 1-23. 1850. Translated into German in *Verhandl. Naturhist. Ver. Preuss. Rheinland und Westphalens*, Vol. VII., pp. 228-254. 1850. This paper is a revision with numerous additions of the article in *La Revue Suisse* of August, 1845, Neuchâtel, entitled: "Notice sur la géographie des animaux." No. **148** of this biography.

1850. — **234.** The natural relations between animals and the elements in which they live. *Amer. Journ. Sc.,* 2d Series, Vol. IX., pp. 369-394. May, 1850. Reprinted in *Ann. Mag. Nat. Hist.,* Vol. VI., pp. 153-179. London, 1850. *Edinb. New Phil. Journ.,* Vol. XLIX., pp. 193-227. Edinburgh, 1850. Translated into French in *Bibl. Univ. Genève Arch. Sc. Phys. et Nat.,* 4e série, Vol. XIX., pp. 15-31. Genève, 1852.

1850. — **235.** Classification of some of the Mollusca Acephala. *Proc. Boston Soc. Nat. Hist.,* Vol. III., p. 301. Boston, June, 1850.

1850. — **236.** On the coloration of animals. *Proc. Amer. Acad. Arts and Sc.,* Vol. II., p. 234. Boston, June, 1850.

1850. — **237.** On the diversified functions of cells. *Proc. Amer. Acad. Arts and Sc.,* Vol. II., p. 236. Boston, July, 1850.

1850. — **238.** On the structure of the egg. *Proc. Amer. Acad. Arts and Sc.,* Vol. II., p. 237. Boston, July, 1850.

1850. — **239.** The diversity of origin of the human races. *Christian Examiner and Religious Miscellany,* Vol. XLIX., pp. 110–145. Boston, July, 1850.

1850. — **240.** On Siluridæ. *Proc. Amer. Acad. Arts and Sc.,* Vol. II., p. 238. Boston, August, 1850.

1850. — **241.** On the scales of the Bonito. *Proc. Amer. Acad. Arts and Sc.,* Vol. II., p. 238. Boston, August, 1850.

1850. — **242.** On the growth of the Egg, prior to the development of the Embryo. *Proc. Amer. Assoc. Adv. Sc.,* Fourth Meeting, New Haven, pp. 18–19. Washington, 1851.

1850. — **243.** On the structure of the mouth in Crustacea. *Proc. Amer. Assoc. Adv. Sc.,* Fourth Meeting, New Haven, pp. 122-123. Washington, 1851.

1850. — **244.** On the relation between coloration and structure in the higher animals. *Proc. Amer. Assoc. Adv. Sc.,* Fourth Meeting, New Haven, p. 194. Washington, 1851.

1850. — **245.** Announcement of an *American Zoölogical Journal* at Cambridge, Massachusetts, "under the direction and editorship of Professor Agassiz," which was never issued. *Amer. Journ. Sc.,* 2d Series, Vol. X., p. 287. September, 1850.

1850. — **246.** A new naked-eyed Medusa, *Rhacostoma Atlanticum. Proc. Boston Soc. Nat. Hist.,* Vol. III., pp. 342–343. Boston, October, 1850.

1850. — **247.** On the pores in the disc of Echinoderms. *Proc. Boston Soc. Nat. Hist.,* Vol. III., pp. 348–349. Boston, October, 1850.

1850. — **248.** On Lamprey Eels (Petromyzontidæ) and their embryonic development and place in the natural history system. Extract from *Agassiz on Lake Superior,* pp. 249–252. *Edinb. New Phil. Journ.,* Vol. XLIX., pp. 242–246. Edinburgh, October, 1850.

1850. — **249.** On the little bodies seen on Hydra. *Proc. Boston Soc. Nat. Hist.,* Vol. III., pp. 354–355. Boston, November, 1850.

1850. — **250.** On the soft parts of American fresh water-Mollusks. *Proc. Boston Soc. Nat. Hist.,* Vol. III., pp. 356–357. Boston, November, 1850.

1850. — **251.** De la classification des animaux dans ses rapports avec leur développement embryonnaire et avec leur histoire paléontologique. *Bibl. Univ. Genève, Arch. Sc. Phys. et Nat.,* 4ᵉ série, Vol. XV., pp. 190–204. Genève, 1850.

1850. — **252.** Classification of Mammalia, Birds, Reptiles, and Fishes from embryonic and paleozoic data. *Edinb. New Phil. Journ.,* Vol. XLIX., pp. 395–398. Edinburgh, 1850.

1850. — **253.** Glacial theory of the erratics and drifts of the New and Old Worlds. *Edinb. New Phil. Journ.,* Vol. XLIX., pp. 97–117. Edinburgh, 1850.

1851. — **254.** The classification of Insects from embryological data. *Smithsonian Contributions to Knowledge,* Vol. II., No. 6. 28 pages. Washington, 1851.

1851. — **255.** Contemplations of God in the Kosmos. *Christian*

Examiner and Religious Miscellany, Vol. L., pp. 1–17. Boston, January, 1851.

1851. — **256.** Extract from the Report of Professor Agassiz to the Superintendent of the Coast Survey, on the Examination of the Florida reefs, keys, and coasts. *Ann. Rep. Supt. Coast Survey during the year ending November*, 1851. Appendix No. 10, pp. 145–160. 8vo. Washington, 1852. Reprinted in the *Report U. S. Coast Survey for the year* 1866. Appendix No. 19, pp. 120–130, 4to. Washington, 1869.

1851. — **257.** On the Florida coral reefs. *Proc. Amer. Acad. Arts and Sc.*, Vol. II., pp. 262–263. Boston, March, 1851.

1851. — **258.** Results of an exploration of the coral reefs of Florida, in connection with the U. S. Coast Survey. *Proc. Amer. Assoc. Adv. Sc.*, Fifth Meeting, Cincinnati, May, 1851, pp. 81–85. Washington, 1851.

1851. — **259.** Report on the vertebrate fossils exhibited to the Association. *Proc. Amer. Assoc. Adv. Sc.*, Fifth Meeting, Cincinnati, May, 1851, pp. 178–180. Washington, 1851.

1851. — **260.** Observations on the Blind Fish of the Mammoth Cave, in a letter to B. Silliman. *Amer. Journ. Sc.*, 2d Series, Vol. XI., pp. 127–128, June, 1851. Reprinted in *Edinb. New Phil. Journ.*, Vol. LI., pp. 254–256, 1851.

1851. — **261.** Shells of New England, by W. Stimpson. 8vo. Boston, 1851. .The author quotes a manuscript of Louis Agassiz on the Naiades of the New England species, pp. 13–15. Isaac Lea has printed the three pages of Stimpson's quotation of Agassiz's Mss., in his "Synopsis of the family of Unionidæ," pp. xix–xx. 4to. Fourth edition. Philadelphia, 1870.

1851. — **262.** Letter to Isaac Lea on Naiades. *Proc. Amer. Phil. Soc.*, Vol. V., p. 219. Philadelphia, September, 1851.

1851. — **263.** Remarks upon the unconformability of the paleozoic formations of the United States. *Proc. Amer. Assoc. Adv. Sc.*, Sixth Meeting, Albany, August, 1851, pp. 254–256. Washington, 1852.

1851.—**264.** On the Mansfield coal formation. *Proc. Amer. Acad. Arts and Sc.*, Vol. II., pp. 270–271. Boston, November, 1851.

1851–1855.—**265.** Grundzüge der Geologie, 1851–1854, and Die Zoologie, 1855. By Agassiz, Gould, and Perty. Stuttgart und Leipzig. An unauthorized publication.

1852.—**266.** Ueber die Gattung unter den nordamerikanischen Najaden. *Wiegmann Archiv, Naturgesch.*, Vol. XVIII., pp. 41–52. Reprinted in part by Isaac Lea in "Synopsis of the family of Unionidæ," pp. xxii–xxiii. Fourth edition. 4to. Philadelphia, 1870.

1852.—**267.** Hugh Miller, author of "Old Red Sandstone" and "Footprints of the Creator," Cambridge, September, 1850. Printed in *The Footprints of the Creator, from the third London edition, with a memoir of the Author by Louis Agassiz*, pp. xi–xxxvii. Boston, 1852.

1852.—**268.** Des relations naturelles qui existent entre les animaux et les milieux dans lesquels ils vivent. *Bibl. Univ. Genève, Arch. Sc. Phys. et Nat.*, 4ᵉ série, Vol. XIX., pp. 15–31. Genève, 1852.

1852.—**269.** Zoölogical notes addressed to J. D. Dana. *Amer. Journ. Sc.*, 2d Series, Vol. XIII., pp. 425–426. May, 1852.

1852.—**270.** Diversity of origin of the human race. *Proc. Amer. Acad. Arts and Sc.*, Vol. III., pp. 7–8. Boston, June, 1852.

1852.—**271.** On the Allantois. *Proc. Amer. Acad. Arts and Sc.*, Vol. III., pp. 15–16. Boston, July, 1852.

1852.—**272.** On organic tissues. *Proc. Amer. Acad. Arts and Sc.*, Vol. III., pp. 21–22. Boston, October, 1852.

1852.—**273.** The earliest larval state of Intestinal Worms. Infusoria. *Edinb. New Phil. Journ.*, Vol. LIII., pp. 314–315. 1852.

1853.—**274.** Sur les Poissons des États-Unis. *Comptes Rendus Acad. sc. France*, Vol. XXXVII., p. 184. Paris, 1853. *L'Institut*, p. 287. Paris, 1853.

1853. — **275.** Family of Cyprinodonts. *Proc. Amer. Acad. Arts and Sc.*, Vol. III., pp. 42–43. Boston, June, 1853.

1853. — **276.** On cell-segmentation. *Proc. Amer. Acad. Arts and Sc.*, Vol. III., pp. 46–47. Boston, June, 1853.

1853. — **277.** Recent researches in a letter addressed to J. D. Dana. *Amer. Journ. Sc.*, 2d Series, Vol. XVI., pp. 134–136, July, 1853.

1853. — **278.** Notices of works on geology. *Amer. Journ. Sc.*, 2d Series, Vol. XVI., pp. 279–283, September, 1853.

1853. — **279.** Notices of works on zoölogy. *Amer. Journ. Sc.*, 2d Series, Vol. XVI., pp. 283–287, September, 1853.

1853. — **280.** On viviparous fishes from California or extraordinary fishes from California constituting a new family. *Amer. Journ. Sc.*, 2d Series, Vol. XVI., pp. 380–390, November, 1853. *Edinb. New Phil. Journ.*, Vol. LVII., pp. 214–228, 1854.

1853. — **281.** On cartilaginous fishes. *Proc. Amer. Acad. Arts and Sc.*, Vol. III., pp. 63–64. Boston, November, 1853.

1853. — **282.** Cestracion from China. *Proc. Amer. Acad. Arts and Sc.*, Vol. III., pp. 65–66. Boston, November, 1853.

1853. — **283.** Age of the new red sandstone of Virginia and North Carolina. *Proc. Amer. Acad. Arts and Sc.*, Vol. III., p. 69. Boston, December, 1853.

1853. — **284.** Footmarks of the Potsdam sandstone. *Proc. Amer. Acad. Arts and Sc.*, Vol. III., p. 70. Boston, December, 1853.

1853. — **285.** Fishes found in the Tennessee River. *Proc. Amer. Acad. Arts and Sc.*, Vol. III., p. 70. Boston, December, 1853.

1854. — **286.** Notice of a collection of Fishes from the southern bend of the Tennessee River in the State of Alabama. *Amer. Journ. Sc.*, 2d Series, Vol. XVII., pp. 297–308, March, 1854; and pp. 353–365, with additional notes on the *Holconoti*, pp. 365–369, May, 1854.

1854. — **287.** The primitive diversity and number of animals in geological times. *Amer. Journ. Sc.*, 2d Series, Vol. XVII., pp. 309–324, May, 1854. *Ann. Mag. Nat. Hist.*, Vol. XIV., pp. 350–366. 1854. *Edinb. New Phil. Journ.*, Vol. LVII., pp. 271–292. Translated into French in *Bibl. Univ. Arch. Sc. Phys. et Nat.*, *Genève*, Vol. XXX., pp. 27–50. 1855.

1854. — **288.** Sketch of the natural provinces of the animal world and their relation to the different types of man (with coloured lithographic tableau and map). Contributed by L. Agassiz, in *Types of Mankind*, by J. C. Nott and G. R. Gliddon, pp. lviii–lxxvi. 4to. Philadelphia, 1854. Reprinted in *Edinb. New Phil. Journ.*, Vol. LVII., pp. 347–363. 1854.

1854. — **289.** Phenomena accompanying the first appearance of a circulating system. *Proc. Amer. Acad. Arts and Sc.*, Vol. III., p. 166. Boston, October, 1854.

1855. — **290.** On the Ichthyological fauna of Western America, or Synopsis of the Ichthyological fauna of the Pacific slope of North America, chiefly from the collections made by the United States Exploring Expedition, under command of Capt. C. Wilkes, with recent additions and comparisons with Eastern types. *Amer. Journ. Sc.*, 2d Series, Vol. XIX., pp. 71–99, January, 1855; and pp. 215–231, March, 1855.

1855. — **291.** Discovery of Viviparous Fish in Louisiana, by Dr. Dowler. *Amer. Journ. Sc.*, 2d Series, Vol. XIX., pp. 133–136. January, 1855. *L'Institut*, Vol. XXIV., p. 164. Paris, 1856. (Poissons Vivipares.)

1855. — **292.** Letter on the Smithsonian Institution, addressed to the Hon. Charles W. Upham. *Amer. Journ. Sc.*, 2d Series, Vol. XIX., pp. 284–287. March, 1855.

1855. — **293.** Lettre à M. Élie de Beaumont sur le développement des êtres ou Transformations embryologiques. *Bull. Soc. géol. France*, 2ᵉ série, Vol. XII., pp. 353–354. Paris, mars, 1855.

1855. — **294.** Classification of Polyps. *Proc. Amer. Acad. Arts and Sc.*, Vol. III., pp. 187–190. Boston, April, 1855.

1855. — **295.** On the so-called footprints of birds in the Connecticut River Sandstone. *Proc. Amer. Acad. Arts and Sc.*, Vol. III., p. 193. Boston, May, 1855.

1855. — **296.** Contributions to the Natural History of the United States of America. *Prospectus.* 4to. Boston, June, 1855. Reprinted in *Amer. Journ. Sc.*, 2d series, Vol. XX., pp. 149-151. July, 1855.

1856. — **297.** Classification in Zoölogy. *Proc. Amer. Acad. Arts and Sc.*, Vol. III., p. 221. Boston, January, 1856.

1856. — **298.** On the Geographical Distribution of Turtles in the United States. *Proc. Boston Soc. Nat. Hist.*, Vol. VI., pp. 6-8. Boston, July, 1856.

1856. — **299.** Ovarian impregnation in Fishes and Chelonians. *Proc. Boston Soc. Nat. Hist.*, Vol. VI., pp. 9-10. Boston, July, 1856.

1856. — **300.** Embryology of a species of shark (*Acantheus Americanus*). *Proc. Boston Soc. Nat. Hist.*, Vol. VI., pp. 37-38. Boston, August, 1856.

1856. — **301.** *Orthagoriscus mola. Proc. Amer. Acad. Arts and Sc.*, Vol. III., p. 319. Boston, August, 1856.

1856. — **302.** On *Cumæ.* In a letter to J. D. Dana. *Amer. Journ. Sc.*, 2d Series, Vol. XXII., pp. 285-286. September, 1856.

1856. — **303.** The class of Fishes, divided into several distinct classes. *Proc. Boston Soc. Nat. Hist.*, Vol. VI., p. 63. Boston, November, 1856.

1856. — **304.** The Glanis of Aristotle. *Proc. Amer. Acad. Arts and Sc.*, Vol. III., pp. 325-333. Boston, November, 1856.

1856. — **305.** On the general characters of orders in the classification of the animal kingdom. *Proc. Amer. Acad. Arts and Sc.*, Vol. III., p. 346. Boston, December, 1856.

1856. — **306.** Notice of the fossil Fishes. *Explorations and Surveys for a railroad route from the Mississippi River to the Pacific Ocean. Report of explorations in California,* by Lieutenant

R. S. Williamson, Vol. V. Appendix, Article I., pp. 313–316. One plate. 4to. Washington, 1856.

Nota bene. — This is the only paper on fossil fishes written by Agassiz, after leaving Europe in 1846.

1857. — **307.** Obituary of Francis C. Gray. *Proc. Amer. Acad. Arts and Sc.*, Vol. III., pp. 347–349. Boston, January, 1857.

1857. — **308.** On the correspondence of different stages of embryonic development with the different stages of geological succession. *Proc. Amer. Acad. Arts and Sc.*, Vol. III., pp. 353–354. Boston, January, 1857.

1857. — **309.** On de Beaumont's theory. *Proc. Amer. Acad. Arts and Sc.*, Vol. III., p. 355. Boston, January, 1857.

1857. — **310.** The order to which Ammonites belong. *Proc. Amer. Acad. Arts and Sc.*, Vol. III., pp. 356–357. Boston, February, 1857.

1857. — **311.** The family of Naiades. *Proc. Amer. Acad. Arts and Sc.*, Vol. III., pp. 378–379. Boston, March, 1857.

1857. — **312.** Nouvelle espèce d'Esoce du Lac Ontario. *L'Institut*, Vol. XXV., p. 128. Paris, 1857.

1857. — **313.** Letter from Professor L. Agassiz, in " Prefatory remarks," by G. R. Gliddon, in *Indigenous Races of the Earth*, pp. xiii–xv. 4to. London and Philadelphia, 1857.

1857. — **314.** Various existing systems of classification of fishes. *Proc. Amer. Acad. Arts and Sc.*, Vol. IV., pp. 8–9. Boston, December, 1857.

1857–1862. — **315.** Contributions to the natural history of the United States of America. 4 vols. 4to. Boston, 1857–1862. Vols. I. and II. were published in 1857; Vol. III., in 1860; Vol. IV., in 1862. Contents: Vol. I., Essay on classification. — North American Testudinata. Vol. II., Embryology of the Turtle. Vol. III., Acalephs in general. — Ctenophoræ. Vol. IV., Discophoræ. — Hydroidæ. — Homologies of the Radiata. Announced in the Prospectus as a work of ten quarto volumes, only four of which were published.

1858. — **316.** What constitutes an individual in natural history? *Proc. Amer. Acad. Arts and Sc.*, Vol. IV., pp. 17–18. Boston, January, 1858.

1858. — **317.** Account of his visit to the reefs of Florida. *Proc. Boston Soc. Nat. Hist.*, Vol. VI., p. 364. Boston, April, 1858.

1858. — **318.** On lasso cells upon living corals. *Proc. Boston Soc. Nat. Hist.*, Vol. VI., p. 367. Boston, April, 1858.

1858. — **319.** Observations upon Corals. *Proc. Boston Soc. Nat. Hist.*, Vol. VI., pp. 373–374. Boston, May, 1858.

1858. — **320.** Observations on the egg-case of Skates. *Proc. Boston Soc. Nat. Hist.*, Vol. VI., pp. 377–378. Boston, May, 1858.

1858. — **321.** Sketch of the labors of the late Professor Johannes Müller. *Proc. Boston Soc. Nat. Hist.*, Vol. VI., pp. 382–383. Boston, June, 1858.

1858. — **322.** A new species of Skate from the Sandwich Islands (*Goniobatis meleagris* Ag.). *Proc. Boston Soc. Nat. Hist.*, Vol. VI., p. 385. Boston, June, 1858.

1858. — **323.** The animals of Millepora are Hydroid Acalephs, and not Polyps. In a letter to J. D. Dana. *Amer. Journ. Sc.*, 2d Series, Vol. XXVI., pp. 140–141. July, 1858. *Bibl. Univ. Genève, Arch. Sc. Phys. et Nat.*, Vol. V., pp. 80–81. 1859.

1858. — **324.** New fishes from Lake Nicaragua, collected by Julius Frœbel. *Proc. Boston Soc. Nat. Hist.*, Vol. VI., pp. 407–408. Boston, October, 1858.

1858. — **325.** Remarks on the Lump-fish (*Discoboli*) from the Florida reefs. *Proc. Boston Soc. Nat. Hist.*, Vol. VI., pp. 411–412. Boston, October, 1858.

1858. — **326.** Some remarks on a catalogue of fishes of Jamaica, by R. Hill of Kingston. *Proc. Boston Soc. Nat. Hist.*, Vol. VI., pp. 414–415. Boston, November, 1858.

1858. — **327.** On some Salmonidæ; The Characini; On the so-called migrations of fishes. *Proc. Boston Soc. Nat. Hist.*, Vol. VI., pp. 418–419. Boston, November, 1858.

1858.—**328.** The classification of Fishes. *Proc. Amer. Acad. Arts and Sc.*, Vol. IV., p. 108. Boston, December, 1858.

1859.—**329.** Similarity between the fauna of Northeastern America and that of Northeastern Asia. *Proc. Amer. Acad. Arts and Sc.*, Vol. IV., pp. 133–134. Boston, January, 1859.

1859.—**330.** On Marcou's "Geology of North America." *Amer. Journ. Sc.*, 2d Series, Vol. XXVII., pp. 134–137. January, 1859. Reprinted in "Reply to the Criticisms of James D. Dana," by Jules Marcou, pp. 26–30. 40 pages, 8vo. Zürich, 1859.

1859.—**331.** On some new Actinoid Polyps of the coast of the United States. *Proc. Boston Soc. Nat. Hist.*, Vol. VII., pp. 23–24. Boston, February, 1859.

1859.—**332.** Origin of Animals. *Proc. Amer. Acad. Arts and Sc.*, Vol. IV., pp. 177–179. Boston, February, 1859.

1859.—**333.** The scientific career of Alexander von Humboldt. *Proc. Amer. Acad. Arts and Sc.*, Vol. IV., pp. 234–247. Boston, May, 1859.

1859.—**334.** Alexander von Humboldt. Eulogy by Professor Agassiz before the American Academy of Arts and Sciences, delivered on the 24th of May. *Amer. Journ. Sc.*, 2d Series, Vol. XXVIII., pp. 96–107. July, 1859.

1859.—**335.** On reversed bivalve shells. *Proc. Boston Soc. Nat. Hist.*, Vol. VII., pp. 166–167. Boston, October, 1859.

1859.—**336.** Discoveries of prehistoric remains on the shores of Lake Neuchâtel. *Proc. Amer. Acad. Arts and Sc.*, Vol. IV., p. 326. Boston, October, 1859.

1859.—**337.** Morphology of the genus *Eurypterus*. *Proc. Amer. Acad. Arts and Sc.*, Vol. IV., p. 353. Boston, December, 1859.

1859.—**338.** The best arrangement of a Zoölogical Museum. *Proc. Boston Soc. Nat. Hist.*, Vol. VII., pp. 191–192. Boston, December, 1859.

1859.—**339.** An essay on classification. An octavo reprint of the "Essay on Classification" contained in Vol. I. of the "Con-

tributions to the Natural History of the United States of America,"
p. 381. London, 1859.

1860. — 340. A communication in opposition to the theory of
origin of species of Mr. Darwin. *Proc. Boston Soc. Nat. Hist.,*
Vol. VII., pp. 231–235. Boston, February, 1860.

1860. — 341. On consecutive faunæ and their corresponding number of geological formations. *Proc. Boston Soc. Nat. Hist.,* Vol. VII.,
pp. 241–245, 250–252. Boston, March, 1860.

1860. — 342. Discussion on the theory of Prof. W. B. Rogers,
of subsidence and denudation of the ocean-floor. *Proc. Boston Soc.
Nat. Hist.,* Vol. VII., pp. 271–275. Boston, April, 1860.

1860. — 343. On the Arctic Sea. *Proc. Amer. Acad. Arts and
Sc.,* Vol. IV., pp. 422–423. Boston, April, 1860.

1860. — 344. Homologies of the Radiata. *Proc. Amer. Acad. Arts
and Sc.,* Vol. IV., p. 441. Boston, May, 1860.

1860. — 345. Individuality and specific differences among Acalephs ; or, Professor Agassiz on the origin of species. Published
from advanced sheets of the third volume of the *Contributions to the
Natural History of the United States. — Amer. Journ. Sc.,* 2d series,
Vol. XXX., pp. 142–154. July, 1860.

1860. — 346. Varieties do not in reality exist as such. *Proc.
Amer. Acad. Arts and Sc.,* Vol. V., p. 72. Boston, October, 1860.

1860. — 347. On the age of some of the sandstones of North
America. *Proc. Boston Soc. Nat. Hist.,* Vol. VII., pp. 356–357.
Boston, 1860.

1860. — 348. On *Mallotus villosus* of Labrador. *Proc. Boston
Soc. Nat. Hist.,* Vol. VII., p. 399. Boston, November, 1860.

1861. — 349. Report of the Director of the Museum of Comparative Zoölogy, for the year 1859, presented to the Board of Trustees
in January, 1860. In *Report of the Trustees of the Museum of
Comparative Zoölogy,* 1861, pp. 33–37. 8vo. Boston, 1861.

1861. — 350. Report of the Director of the Museum of Comparative Zoölogy, for the year 1860. January 30, 1861. In *Report of*

the Trustees of the Museum, in 1861, pp. 43-49. 8vo. Boston, 1861.

1861. — **351.** Discussion on the primordial fauna. *Proc. Boston Soc. Nat. Hist.*, Vol. VIII., pp. 58-59. Boston, January, 1861.

1861. — **352.** Some remarks on the circumscription of animals in the ocean. *Proc. Boston Soc. Nat. Hist.*, Vol. VIII., p. 60. Boston, January, 1861.

1861. — **353.** Observations on the rate of increase and other characters of fresh-water shells, Unios. *Proc. Boston Soc. Nat. Hist.*, Vol. VIII., pp. 100-102. Boston, February, 1861.

1861. — **354.** Perforation in rocks made by the *Saxicava rugosa*, a bivalve shell. *Proc. Boston Soc. Nat. Hist.*, Vol. VIII., p. 102. Boston, February, 1861.

1861. — **355.** Two individual corals developed from one base. *Proc. Boston Soc. Nat. Hist.*, Vol. VIII., p. 104. Boston, February, 1861.

1861. — **356.** Pressure on living star-fishes at great depths. *Proc. Boston Soc. Nat. Hist.*, Vol. VIII., p. 104. Boston, February, 1861.

1861. — **357.** On the homologies of Echinoderms. *Proc. Boston Soc. Nat. Hist.*, Vol. VIII., pp. 235-238. Boston, November, 1861.

1861. — **358.** Remarks on bilateral symmetry and laterality in mollusks. *Proc. Boston Soc. Nat. Hist.*, Vol. VIII., p. 279. Boston, November, 1861.

1862. — **359.** Third Annual Report of the Museum of Comparative Zoölogy, October, 1861. In *Annual Report of the Trustees of the Museum of Comparative Zoölogy, together with the Report of the Director*, 1862, pp. 5-17. 8vo. Boston, 1862.

1862. — **360.** Directions for collecting objects of natural history, by L. Agassiz, Director of the Museum of Comparative Zoölogy. 8 p. 8vo. No date, no place of publication. Cambridge, 1862.

1862. — **361.** Highly interesting discovery of new Sauroid remains. In a letter to B. Silliman. *Amer. Journ. Sc.*, 2d series, Vol. XXXIII., p. 138. January, 1862.

1862. — **362.** The structure of animal life. Six lectures delivered at the Brooklyn Academy of Music, in January and February, 1862. New York. A fourth edition was issued at New York, in 1886.

1862. — **363.** On homologies of Brachiopoda. *Proc. Boston Soc. Nat. Hist.*, Vol. IX., pp. 68–69. Boston, May, 1862.

1862. — **364.** On the Megatheroids. *Proc. Boston Soc. Nat. Hist.*, Vol. IX., pp. 101–102. Boston, June, 1862.

1862. — **365.** On development of *Rana temporaria*. *Proc. Boston Soc. Nat. Hist.*, Vol. IX., p. 174. Boston, October, 1862.

1862. — **366.** On the subdivisions of Tertiary strata. *Proc. Boston Soc. Nat. Hist.*, Vol. IX., p. 174. Boston, October, 1862.

1862. — **367.** Differences among the faunæ of fossils. *Proc. Amer. Acad. Arts and Sc.*, Vol. VI., p. 81. Boston, October, 1862.

1862. — **368.** On geographical distribution of the fresh-water fishes. *Proc. Boston Soc. Nat. Hist.*, Vol. IX., p. 178. Boston, November, 1862.

1863. — **369.** Methods of study in natural history. 12mo. Boston, 1863.

Nota bene. — Appeared first in serial form in the *Atlantic Monthly*, Vol. IX., 1861, pp. 1–13, 214–222, 327–337, 446–460, 570–578, 754–762; and Vol. X., 1862, pp. 87–98, 325–336, 571–580; under the following titles : General sketch of the early progress in natural history. Nomenclature and classification. Categories in classification. Classification and creation — Different views respecting orders. Gradation among animals. Analogous types. Family characteristics. The characters of genera. Species and breeds. Formation of coral reefs. Age of coral reefs as showing permanence of species. Homologies. Alternate generations. The ovarian egg. The closing chapter, Embryology and Classification, did not appear in the *Atlantic Monthly*.

This is the most popular volume published by Agassiz. A nineteenth edition was issued in 1889.

1863. — **370.** Fourth Annual Report of the Museum of Comparative Zoölogy, October, 1862. In *Annual Report of the Trustees of the Museum*, etc., 1862; pp. 5–13. 8vo. Boston, 1863.

1863.— **371**. On the enigmatic fossil of Solenhofen. *Proc. Boston Soc. Nat. Hist.*, Vol. IX., p. 191. Boston, January, 1863.

1863. — **372**. Geographical distribution of Echini. *Proc. Boston Soc. Nat. Hist.*, Vol. IX., pp. 191–192. Boston, January, 1863.

1863. — **373**. On the natural attitude of the Megatherium. *Proc. Boston Soc. Nat. Hist.*, Vol. IX., p. 193. Boston, January, 1863.

1863.— **374**. On the young of fishes. *Proc. Boston Soc. Nat. Hist.*, Vol. IX., p. 326. Boston, October, 1863.

1864. —**375**. Fifth Annual Report of the Director of the Museum of Comparative Zoölogy. In *Annual Report of the Trustees of the Museum*, etc., 1863, pp. 6–18. 8vo. Boston, 1864.

1865. — **376**. Sixth Annual Report of the Director of the Museum of Comparative Zoölogy, at Harvard College, in Cambridge, Massachusetts. In *Annual Report of the Trustees of the Museum*, etc., 1864, pp. 7–17. 8vo. Boston, 1865.

1865. — **377**. Métamorphoses subies par certains Poissons avant de prendre la forme propre à l'adulte. *Comptes Rendus Acad. sc. France*, Vol. LX., pp. 152–153. Paris, 1865. *Ann. sc. nat. Zoologie*, 5° série, Vol. III., pp. 55–58. Paris, 1865. *Ann. Mag. Nat. Hist.*, 3d series, Vol. XVI., pp. 69–70. London, 1865.

1865–1866. — **378**. Lettres relatives à la faune Ichthyologique de l'Amazone. *Ann. sc. nat. Zoologie*, 5° série, Vol. IV., pp. 382–383, Paris, 1865 ; Vol. V., pp. 226–228, 309–311, Paris, 1866.

1866. — **379**. Geological sketches. First series, 12mo, Boston, 1866. Reprinted in 1870.

Nota bene. — Appeared first in serial form in the *Atlantic Monthly*, Vol. XI., 1863, pp. 373–382, 460–471, 615–625, 742–756 ; and Vol. XII., pp. 72–81, 212–224, 333–342, 568–576, 751–767 ; and Vol. XIII., 1864, pp. 56–65 ; under the following titles : America the Old World. The Silurian beach. The fern forests of the Carboniferous period. Mountains and their origin. The growth of continents. The geological middle age. The Tertiary age and its characteristic animals. The formation of glaciers. Internal structure and progression of glaciers. External appearance of glaciers.

1866. — **380**. Çonversações scientificas sobre o Amazonas feitas na sala do externato do Collegio de Pedro II., durante o mez de Maio de 1866. Translated into Portuguese by Féiix Vogeli. 71 pages, 8vo. Rio de Janeiro, 1866. A French translation of a part has appeared under the title, "Aperçu du cours de l'Amazone d'après le Professeur Agassiz" par la *Rédaction* du Bulletin de la Société de Géographie de Paris (Charles Maunoir, Secrétaire-général). *Bull. Soc. Géographie, Paris*, Vol. XII., pp. 433-457. Paris, décembre, 1866. Reprinted under the title, "Bassin de l'Amazone" in *Mém. et Bull. Soc. Géogr. de Genève*, Vol. VII., pp. 150-196. Genève, 1868.

1866. — **381**. Lettre à Marcou sur la géologie de la vallée de l'Amazone occupée par un lœss, avec des remarques de Jules Marcou. *Bull. Soc. géol. France*, 2 série, Vol. XXIV., pp. 109-110. Paris, décembre, 1866. *Leonhard und Bronn, Neues Jahrb.*, Vol. XXXVIII., pp. 180-181. 1867.

1867. — **382**. Report on use of a new hall in the Smithsonian. *Report Smithsonian Institution for* 1867, pp. 109-111. Washington, 1868.

1867. — **383**. Annual Report of the Director of the Museum of Comparative Zoölogy on resuming his duties in 1866. In *Annual Report of the Trustees of the Museum*, etc., 1866, p. 4, with a "Special Report of the Director," pp. 8-17. 8vo. Boston, 1867.

1867. — **384**. Observations géologiques faites dans la vallée de l'Amazone. *Comptes Rendus Acad. sc. France*, Vol. LXIV., pp. 1269-1270. Paris, 1867. *Leonhard und Bronn, Neues Jahrb.*, Vol. XXXVIII., pp. 676-680. 1867.

1867. — **385**. Remarks upon the antiquity of man. *Proc. Boston Soc. Nat. Hist.*, Vol. XI., pp. 304-305. Boston, October, 1867.

1867. — **386**. On phyllotaxis. *Proc. Boston Soc. Nat. Hist.*, Vol. XI., pp. 315-316. Boston, November, 1867.

1867. — **387**. Examination of the skulls of the American bison and the European aurochs. *Proc. Boston Soc. Nat. Hist.*, Vol. XI., pp. 316-318. Boston, November, 1867.

1867. — **388.** A Cetacean new to America. *Proc. Boston Soc. Nat. Hist.*, Vol. XI., p. 318. Boston, November, 1867.

1867. — **389.** Remarks on the Taconic system. *Proc. Boston Soc. Nat. Hist.*, Vol. XI., pp. 353–354. Boston, December, 1867.

1867. — **390.** On the classification of the Siluroids. *Proc. Boston Soc. Nat. Hist.*, Vol. XI., p. 354. Boston, December, 1867.

1868. — **391.** Report of the Director of the Museum. In *Annual Report of the Trustees of the Museum of Comparative Zoölogy at Harvard College, in Cambridge, together with the Report of the Director,* 1867, pp. 4–12. 8vo. Boston, 1868.

1868. — **392.** Sur la géologie de l'Amazone, par MM. Agassiz et Coutinho; notice rédigée et communiquée par M. Jules Marcou. *Bull. Soc. géol. France,* 2ᵉ série, Vol. XXV., pp. 685–691. Paris, mai, 1868.

1868. — **393.** A Journey in Brazil by Professor and Mrs. Louis Agassiz. 540 pages. 8vo. Boston, 1868.

1869. — **394.** Voyage au Brésil par Mme. et M. Louis Agassiz, traduit de l'Anglais par Félix Vogeli. Avec additions et plus de gravures et de cartes. 532 pages. 8vo. Paris, 1869. This is more complete than the English edition of 1868.

In 1872 an abridged edition was published at Paris, under the title, " Voyage au Brésil de Louis Agassiz, abrégé sur la traduction de F. Vogeli." 12mo. Carte et 16 gravures.

1869. — **395.** Report of the Director of the Museum of Comparative Zoölogy, for the year 1868. In *Annual Report of the Trustees of the Museum,* etc., 1868, pp. 4–12. 8vo. Boston, 1869.

1869. — **396.** De l'espèce et de sa classification en Zoologie. Traduit de l'anglais par Félix Vogeli. Edition française revue et augmentée par l'auteur de *l'Essay on Classification.* 8vo. Paris, 1869.

1869. — **397.** Principes rationnels de la classification Zoologique. *Revue des Cours Scientifiques,* Vol. VI., pp. 146–165. Paris, 1869. Extrait de la traduction française, "An essay on classification."

1869. — **398.** Nature et définition des espèces. *Revue des Cours*

Scientifiques, Vol. VI., pp. 166–169. Paris, 1869. Extrait du volume, traduit en français, "An essay on classification."

1869. — **399**. Ordre d'apparition des caractères zoologiques pendant la vie embryonnaire. *Revue des Cours Scientifiques*, Vol. VI., pp. 169–171. Paris, 1869. Extrait du volume traduit en français, " An essay on classification."

1869. — **400**. Address delivered on the centennial anniversary of the birth of Alexander von Humboldt, under the auspices of the Boston Society of Natural History. 58 pages, 8vo. Boston, 1869. The leading newspapers of Boston, New York, etc., reprinted Agassiz's address in full.

1869. — **401**. Report upon deep-sea dredgings in the Gulf Stream, during the third cruise of the United States steamer *Bibb*, addressed to Professor Benjamin Pierce, superintendent United States Coast Survey. *Bull. Mus. Comp. Zoöl.*, Vol. I., pp. 363–386. Cambridge, November, 1869.

1870. — **402**. Report of the Director of the Museum of Comparative Zoölogy, for the year 1869. In *Annual Report of the Trustees of the Museum*, etc., 1869, pp. 4–11. 8vo. Boston, 1870.

1870. — **403**. On the former existence of local glaciers in the White Mountains. *Proc. Amer. Assoc. Adv. Sc.*, Nineteenth Meeting, Troy, New York, Vol. XIX., pp. 161–167. Cambridge, 1871. *Amer. Naturalist*, Vol. IV., pp. 550–558. 1871.

1871. — **404**. Report of the Director of the Museum of Comparative Zoölogy, for the year 1870. In *Annual Report of the Trustees of the Museum*, etc., 1870, pp. 4–8. 8vo. Boston, 1871.

1871. — **405**. Eulogy of Dr. J. E. Holbrook. *Proc. Boston Soc. Nat. Hist.*, Vol. XIV., pp. 347–351. Boston, October, 1871.

1871. — **406**. Observations on a set of boulders in Berkshire County and Wachusett range, Massachusetts. *Proc. Boston Soc. Nat. Hist.*, Vol. XIV., pp. 385–386. Boston, October, 1871.

1871. — **407**. Mode of Copulation among the Selachians. *Proc. Boston Soc. Nat. Hist.*, Vol. XIV., pp. 339–341. Boston, October, 1871.

1871. — **408.** Letter concerning deep-sea dredging, addressed to Professor Benjamin Pierce. *Bull. Mus. Comp. Zoöl.*, Vol. III., pp. 49–53. Cambridge, December, 1871. *Ann. Mag. Nat. Hist.*, Vol. IX., pp. 169–174. London, 1872.

1872. — **409.** Report of the Director of the Museum of Comparative Zoölogy, for the year 1871. In *Annual Report of the Trustees of the Museum*, etc., 1871, pp. 4–8. 8vo. Boston, 1872.

1872. — **410.** Fish-nest (of *Chironectes Pictus*) in the seaweed of the Sargasso Sea. *Amer. Journ. Sc.*, 3d Series, Vol. III., pp. 154–156. 1872. *Ann. Mag. Nat. Hist.*, 4th Series, Vol. IX., pp. 243–245. London, 1872. *Bulletin Soc. sc. nat. Neuchâtel*, Vol. IX., pp. 165–169. Neuchâtel, 1873.

1872. — **411.** Agassiz's deep-sea explorations. *More about the trilobites.* In a letter to Professor Pierce, published in *The New York Tribune*, and reprinted in *The Canadian Naturalist*, Vol. VI., New Series, pp. 358–361. Montreal, 1872.

1872. — **412.** Glacial action in Fuegia and Patagonia. Abstract of a letter by Professor Agassiz of the Hassler Expedition, addressed to Prof. B. Pierce, dated Talcahuana, April 27. *Amer. Journ. Sc.*, 3d Series, Vol. IV., pp. 135–136. August, 1872.

1872. — **413.** Address to the California Academy of Science, in Response to an Introduction. *Proc. California Acad. Sc.*, Vol. IV., p. 253, Sept. 2, 1872. San Francisco, 1872.

1872. — **414.** Remarks on results of the Hassler Expedition. *Proc. California Acad. Sc.*, Vol. IV., pp. 257–258, Sept. 2. San Francisco, 1872. *Mining and Scientific Press*, Vol. XXV., p. 153, Sept. 7. San Francisco, 1872.

1872. — **415.** A lecture on the natural history of the animal kingdom. *Mining and Scientific Press*, Vol. XXV., pp. 262–265. San Francisco, October, 1872. *Overland Monthly*, Vol. IX., pp. 461–466. San Francisco, October, 1872.

1872. — **416.** Sketch of a voyage from Boston to San Francisco. Professor Agassiz's Narrative. *Smithsonian Report for* 1872. Appendix to the *Journal of Proceedings of the Board of Regents,* pp. 87–92. Washington, 1873. Also in *Misc. Coll. Smithsonian Institution*, Vol. XVIII., pp. 394–400. Washington, 1872.

1873. — **417.** Voyage d'exploration scientifique dans l'Atlantique et l'Amérique du Sud. *Revue des Cours Scientifiques*, 2ᵉ série, Vol. IV., pp. 1077–1093. Paris, 1873.

1873. — **418.** Structure and growth of domesticated animals. *Amer. Naturalist*, Vol. VII., pp. 641–657. Salem, 1873. *Twentieth Annual Report of the Mass. Board of Agriculture.* Boston, December 3, 1872.

Posthumous Publications.

1874. — **419.** Evolution and permanence of type. *Atlantic Monthly*, Vol. XXXIII., pp. 92–101. Boston, January, 1874.

1874. — **420.** The Darwinian theory. Fac-simile of a letter sent to James A. Parsons, in reply to an inquiry as to Agassiz's views. *Scientific American*, Vol. XXX., p. 85. February 7, 1874.

1874. — **421.** Three different modes of teething among Selachians. *Amer. Naturalist*, Vol. VIII., pp. 129–135. Salem, March, 1874.

1874. — **422.** Two letters addressed to Alexander Murray and Jules Marcou on Gigantic Cuttle-Fishes of Newfoundland. *Amer. Naturalist*, Vol. VIII., pp. 226–227. Salem, April, 1874. The letter to Alexander Murray is reprinted in *Maritime Monthly Mag.*, Vol. III., p. 207. Saint John, New Brunswick, March, 1874.

1874. — **423.** The organization and progress of the Anderson School of Natural History at Penikese Island. *Report of the Trustees for* 1873. Contains several letters, addresses, and a circular by Louis Agassiz. 20 pages and 3 plates, 8vo. Cambridge, 1874.

1876. — **424.** Geological sketches. 2d Series, 12mo. Boston, 1876.

Nota bene. — Appeared first in serial form in the *Atlantic Monthly*, Vol. XIII., 1864, pp. 224–232, 723–736; Vol. XVIII., 1866, pp. 49–60, 159–169; and Vol. XIX., 1867, pp. 211–220, 281--

287; under the following titles: Glacial Period, The Parallel Roads of Glen Roy in Scotland, Ice Period in America, Glacial Phenomena in Maine, Physical History of the Valley of the Amazons. Edited by Mrs. E. C. Agassiz.

The paper "Glacial Phenomena in Maine," published in the *Atlantic Monthly*, Vol. XIX., pp. 211–220 and 281–287, February and March, 1867, has been translated into French, and published in *Bibl. Univ. Genève, Arch. Sc. Phys. et Nat.*, Vol. XXVIII., pp. 319–352. Genève, 1867.

1880. — 425. Report on the Florida reefs, by Louis Agassiz, accompanied by illustrations of Florida corals, from drawings by A. Sonrel, Burkhardt, Alexander Agassiz, and Roetter; with an explanation of the plates by L. F. de Pourtalès. 21 pages and 23 plates. May, 1880. *Mem. Mus. Comp. Zoöl.*, Vol. VII., No. 1. Cambridge, 1882.

INDEX.

Aar, Glacier of, i. 147, 163, 175, 187, 195, 203, 219, 250, 251, 263.
Academy of Natural Sciences, Philadelphia, ii. 27.
Agasse, M., publisher in Paris, i. 3.
Agassiz, family origin, i. 1; different branches of the family, i. 2; coat of arms, i. 2; name in Arabic, Mauresque, and Saracenic language, i. 3; name in Italy, i. 3; an Agassiz married to a French Huguenot, i. 4; family connections in the Cevennes and Provence, i. 4; descendants of French Huguenots mere tradition, i. 4; family features Swiss and Jurassic, i. 4.
Agassiz, Rodolphe Benjamin Louis, father of Louis, i. 5; pastor at Motier, i. 5; leaves St. Imier, i. 6; his children, i. 7; as a teacher, i. 8.
Agassiz, Auguste, at the College of Bienne, i. 10.
Agassiz, Mrs. Rose, i. 14.
Agassiz, Mrs. Cecile, as an artist, i. 57; dislikes Neuchâtel, i. 58; an excellent and careful mother, i. 66; in poor health, i. 246; joins her family at Carlsruhe, i. 246; her death, ii. 18.
Agassiz, Alexander, i. 232, 262, ii. 51, 61, 88, 137, 192.
Agassiz, Jean Louis Rodolphe.
    VOL. I. Born at Motier, 7; takes naturally to water, 8; the best pupil of his father, 8; his passion for collecting objects of natural history, 9; goes to the College of Bienne, 9; his capacity for languages, 9; geography his favourite study, 10; walks to and from Motier and Bienne, 10; vacations spent at his grandfather's, Dr. Mayor, 11; country life at Cudrefin, 11; as a sportsman, 12; enters the

commercial house of his uncle, 13; at the College of Lausanne, 13; resolves to become a naturalist, 13; asks permission to study medicine, 14; as an admirer of the fair sex, 14; goes to the University of Zurich, 14; studies ornithology, 14; student life at Zurich, 15; his friendship for Arnold Escher, 15; his motto "First at work, and first at play," 16; his constitution, 16; goes to the University of Heidelberg, 16; becomes acquainted with Alexander Braun and Karl Schimper, 16; visits Carlsruhe, 16; sickness at Heidelberg, 18; explores the environs of Orbe, 38; writes his first essay in natural history at Orbe, 18; his first artists, the two "Ceciles," 19; joins Braun at Carlsruhe, 20; portrait by Miss Cecile Braun, 20; goes to Munich, 21; visits the Royal Museum at Stuttgart, 21; visits Esslingen, 21; visit to Ferdinand von Hochstetter, 21; at Munich, with Braun and Schimper, 21; at the house of Dollinger, 22; becomes Germanized, 22; *Erwiederung auf Dr. Karl Schimper's Angriffe*, 22; as a swordsman, 22; his pleasure in fencing, 23; challenges the German Club, 23; confines himself to work on fishes, 24; his yearly allowance, 24; student life at Munich, 24; his happy life at Munich, 25; his acquaintance courted by all, 25; doctor of medicine, 26; doctor of philosophy from the University of Erlangen, 26; Martius secures his services on the fishes of Brazil, 27; receives his degree of doctor of medicine, 27; his fishes of Brazil attracts the attention of Cuvier, 28;

.

Printed in the United States
By Bookmasters